Electrical Soldering

by

Louis M. Dezettel

A Revision of

abc's of Electrical Soldering

by

Louis M. Dezettel

Howard W. Sams & Co., Inc.
4300 WEST 62ND ST. INDIANAPOLIS, INDIANA 46268 USA

SECOND EDITION
SECOND PRINTING—1979

International Standard Book Number: 0-672-21411-3
Library of Congress Catalog Card Number: 76-42878

Printed in the United States of America.

Preface

Electrical soldering is both a science and a skill. As a science, it deals with the metallurgy of metal alloys and the chemistry of fluxes. As a skill, it calls on know-how and experience in the actual hand soldering operation to make the best-quality soldered connections.

This book deals with both the science and the skill of soldering. The science of soldering is covered only enough to help the user of solder and soldering irons better understand his equipment, and thus more easily develop his skill. Most of the information about the metallurgy and chemistry of the products used in soldering comes out of the laboratories of the solder and flux processors. It is the processors' responsibility to offer to a highly competitive market the products that will do the best job at an economical price. The processors supply the products to the user, and it is up to the user to understand enough about those products to make a wise choice, based on his particular needs, and to put them to proper use.

Soldering skill is not hard to acquire. The fundamentals of soldering are really quite simple. Once experience is added to those fundamentals, skill comes naturally, and good solder connections are automatically made.

This book attempts to cover the subject of electrical soldering for three classes of solderers: the beginner, the technician, and the manufacturer. The electronics experimenter-beginner who has never soldered before, or who has done very little soldering, must get the right start. It is important for him to form the right soldering habits. With the information in this book, he can begin making good soldering connections almost at once. For someone who has never soldered before, the greatest obstacle to overcome is lack of confidence. The beginner seems to fear that he is going to damage the connection with the heat and so he uses too little. The best way to overcome this fear is to understand what is involved in the skill of soldering and to practice using the correct methods.

Radio and television servicemen and laboratory technicians are men of experience in soldering. Soldering practices become habits. But are the habits good ones? Does the technician really understand soldering? This book is an excellent review, either to assure the technician that he has learned his skills properly or to tell him how he can change his habits to do a better job.

Many manufacturers are entering the field of electrical soldering for the first time. The manufacturer should consult his suppliers of solder and fluxes and printed-circuit boards and let them recommend products to meet his needs. But before the manufacturer meets with representatives from suppliers, he should have adequate knowledge of soldering so that he can be sure that the products being recommended to him are really the best for his operation. There is enough information in this book to prepare the manufacturer for such a meeting.

LOUIS M. DEZETTEL

Contents

CHAPTER 1

WHAT SOLDERING IS AND DOES 7
Joining Metals — What Is a Metal? — Soldering — Representative Soldering Irons

CHAPTER 2

THE SOLDER ALLOY 17
Soft Solder — Tin-Lead Temperature Graph — Solder Additives — Solder Qualities — The Forms of Solder — Aluminum Soldering

CHAPTER 3

FLUXES 24
Organic and Inorganic Fluxes — Cleanliness — Flux Cores — Non-core Soldering

CHAPTER 4

SOLDERING IRONS 30
Soldering-Iron Construction — Pencil-Type Irons — Transformer-Isolated Irons — Iron Costs — Construction Details — Heat Control — Tips — Care of Irons — Car-Battery-Operated Iron

CHAPTER 5

INSTANT-HEAT GUNS 46
How It Works — Self-Regulation — Manufacturers of Instant-Heat Guns — Accessory Tips — The Compromise Gun

CHAPTER 6

THE CORDLESS SOLDERING IRON 52
Advantages of the Cordless Iron — A Different Soldering Tip — The Charger — Battery Care — The Weller Cordless Iron — Ungar Cordless Iron and Stand — The Cordless Wahl With Drill Attachment — The Wen Cordless Gun — The Lenk Cordless Iron

CHAPTER 7

HOW TO SOLDER 61
Soldering Superiority — The Right Iron and Tip — Solder and Flux — Cleanliness — Wire Stripping — Pretinning Leads — Soldering for Beginners — Experienced Solderers — Printed-Circuit Boards — Switch Terminals — Soldering to Chassis — Semiconductors — Cable Connectors — Shielded Cables — Wire Splices — Inspection — Tacking — Soldering Aids

CHAPTER 8

PRODUCTION SOLDERING 83
Hand Soldering — Voltage Leakage — Soldering Iron Details — Resistance Soldering — Soldering Preforms — The Soldering Pot — Dip Soldering — Wave Soldering — Board Treatment — Solder Resist — Solderability Tests — Stands — Temperature-Controlled Stands

CHAPTER 9

MISCELLANEOUS SOLDERING AND CONNECTING METHODS . . . 114
The Bottled-Gas Torch — The Friction Wheel — Solderless Connections — Other Methods — Special Metals

CHAPTER 10

HOW TO MAKE SOLDER REPAIRS 124
Soldering/Desoldering Aids — Special Desoldering Tips — Printed-Circuit Board Wiring Repairs — Vacuum Desoldering Irons — Wick Solder Removers

CHAPTER 11

BURNS AND THEIR TREATMENT 137
Precautions — Classification of Burns — Treatment of Burns

INDEX 141

1

What Soldering Is and Does

Soldering as an art is not new. It dates back some 4000 years. Soldering as an exact science is new. It dates back to only about the turn of the century.

Evidence of silver soldering has been found in the vases of early cultures as far back as 2000 to 3000 B.C. "Soft" soldering, using lead and alloys of lead, has been used in the crafts for centuries. In the days of the Roman Empire, water pipes were made of lead. Lead was rolled into long flat plates and shaped into pipes; then the edges were soldered. Our modern word *plumbing* comes from the Latin *plumbum*, the word for lead. The chemical symbol for lead is *Pb*.

The more recent use of solder as an alloy of tin and lead began with the tin can, at about the turn of the century. At that time, the alloying of precise amounts of tin and lead made it possible to heat-join the metals of tin cans, and thus made it both economical and practical to use tin cans as containers for food.

The more recent rapid increase in the knowledge of solder alloys and fluxes parallels the growth of electronics. Soldering is now the most popular method of making good electrical connections. Today millions of solder connections are made in space exploration projects, without a single failure.

JOINING METALS

"Soldering" means the joining of metals by heat for the purpose of making something structural, or producing a continuous and permanent path for the flow of electricity. Metals for structural assembly can be joined by soldering, brazing, or welding. *Soldering* entails using a low-melting-point alloy and heating it to a liquid state so that it flows onto the base metal being joined and alloys with it at the surface. (In resistance soldering, the connection itself is heated—not the alloy—by means of a heavy electric current passed through the connection.) *Brazing* is like soldering but is done at higher temperatures and with different alloys and fluxes. *Welding* is done at still higher temperatures, where the base metals themselves are brought to a liquid state and fused together. No separate alloy material is used in welding.

The purpose of this book is to teach you to make electrical connections by soldering. Welding is sometimes used, for example, to weld railroad rails end to end to give them strength and electrical continuity. Brazing is used on occasion for electrical continuity in heavy-current applications. In high-current dc generators, heavy rotor coil wires may be brazed to commutator segments. However, by far the most popular system for making electrical connections is soldering. Its popularity is due to the low amount of heat needed to melt the solder alloy (under 800°F), thus making tools and techniques simple in comparison with the other two methods.

Welding, brazing, and soldering are listed here in order of decreasing strength, but the strength of a good soldering job is sufficient for nearly all electrical connections.

WHAT IS A METAL?

The atomic structure of metals is rather different from that of other elements. A metal atom has a number of free electrons in its outer shell. These free electrons easily interchange with the free electrons of atoms adjacent to them (Fig. 1-1). The electrical

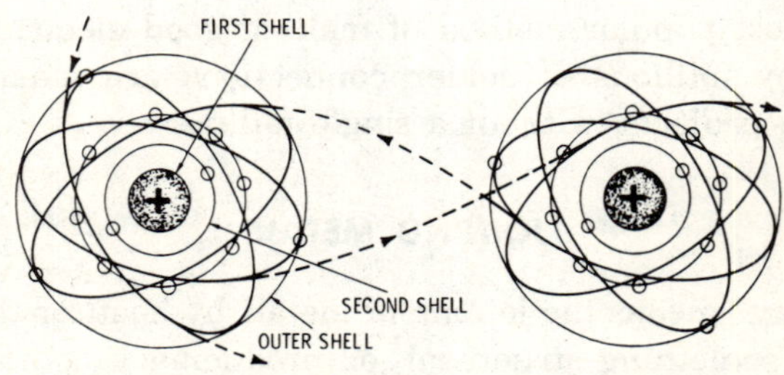

Fig. 1-1. Free electrons in metals interchange easily.

attraction of the negative electrons to positive ions results in a strong adhesion, which helps to make metals solid. The movement of free electrons is also what gives metals the ability to carry electrical current and to have good heat conductivity. A further result is to produce many voids in the crystal-lattice outer structure of each metal molecule. These voids make it possible to make *alloys* of metals, which are simply mixtures of more than one metal (Fig. 1-2).

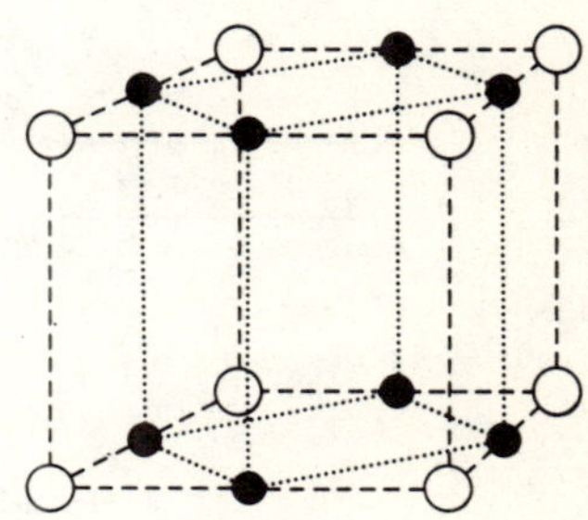

Fig. 1-2. Voids in lattice structure permit formation of alloys (grossly simplified).

In an alloy, different metals can be combined in any proportion without the properties of each metal being affected too much, provided the sizes of the molecules are similar. This combining of metals is something like mixing water and alcohol, which will stay in solution when mixed in any proportion and, if desired, can be separated by distillation.

SOLDERING

Solder is an alloy of tin and lead (with other metals sometimes added in small quantities for special purposes) that melts into a liquid state at very low temperatures compared with other metals. When in its hot molten state, it will, on contact with another metal, transfer its thermal activity by convection to the surface of the other metal (called the *base metal*). The solder alloy will combine with the base metal at the interface of the two (Fig. 1-3). The edges of the crystal lattice structure overlap each other and join to form a solid that is mechanically and electrically continuous. To visualize it, you might compare it to the action of cements on two pieces of paper or wood. The liquid cement flows into the porous outer surfaces and hooks the two pieces of paper or wood together when the cement dries. In soldering, the adhesion is not mechanical but chemical.

Molten solder dropped onto a nonmetallic surface, or the surface of metal that is not clean, will form a nearly round ball, like water on a nonporous surface. Surface tension keeps the solder in the form of a ball; the effect of gravity causes it to flatten. Molten solder dropped onto the *clean* (more about this later) surface of certain metals will develop a capillary attraction that overcomes

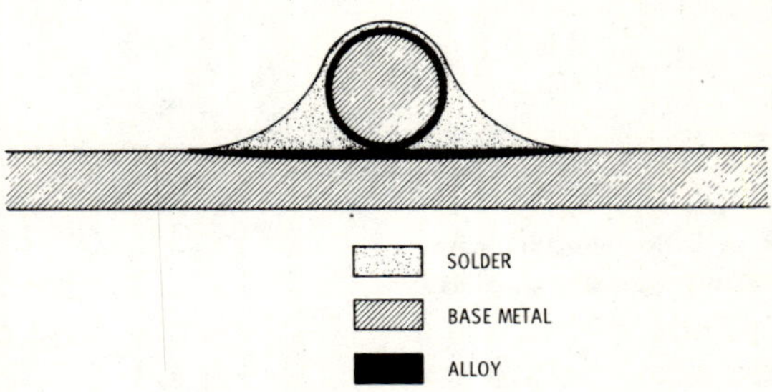

Fig. 1-3. Process of soldering.

the surface tension, and will flow along the surface of the metal (Fig. 1-4), like a drop of water on a clean paper towel. This is known as *wetting* the base metal, and the secret of good soldering is the ability to properly wet the base metal with the solder. Wetting is an intermetallic solution action between the solder and the base metal.

The actual depth of the intermixing of the solder with the base metal is very shallow, probably not even one molecule thick. But it is sufficient to make a good bond.

(A) Onto nonmetallic or unclean surface. (B) Onto clean metallic surface.

Fig. 1-4. Molten solder dropped onto various surfaces.

In making electrical connections, soldering should be considered an important supplement to the connection. A firm mechanical connection is first made between two wires or between wires and a terminal. The added solder bond strengthens the connection mechanically, the flux removes the oxide barrier at the interfaces, and the solder coating protects the surfaces of the base metals at the junction from further surface corrosion.

A Flux Is Also Needed

Capillary attraction between the solder and the base metal will not take place unless the base metal is clean. "Clean" here does not mean just clean in appearance, but clean of all invisible surface oxides. Under practical conditions, even bright copper will have a thin oxide film. A *flux* is used with the solder to remove the oxide, and thus permit good solder wetting.

What Can Be Soldered?

Only certain metals are atomically structured to accept wetting by a tin-lead alloy solder. Of these, copper is the one we deal with most often. Others are silver, iron and steel, zinc, nickel, gold, platinum, and palladium. Some metals are solderable with special alloys and fluxes. Aluminum, stainless steel, and magnesium are among them. A few metals are almost impossible to solder. Among these are chromium and cobalt.

Tin and lead were not mentioned above because, while wettable with solder, they would melt with the heat of the soldering operation. An accessory solder alloy is not needed to solder to tin or lead.

While the metallurgical science behind soldering is interesting, it is not necessary to know it to make good solder connections. Anyone can learn to solder well. The manufacturers of solder alloys and of soldering irons have done the research and have come up with products anyone can use successfully. The background in this book will give the serious reader a better understanding of soldering. It is of special benefit to the technician who does a lot of soldering, and to industry which may be doing constant soldering operations in production. But a novice, too, can learn much, and with practice he can make good connections for his electronic projects.

Courtesy Wall-Lenk Manufacturing Co.

Fig. 1-5. An example of the new trend in soldering irons. A single handle accommodates several interchangeable heating elements and tip sizes.

There is more to soldering than what has been said so far in this chapter. Cleanliness of metal surfaces is of prime importance. This is ensured by precleaning and the use of rosin-base fluxes. The tin-to-lead ratios affect the connection to a degree, and the temperature of the heat applied to the connection is of even greater importance. This is where the soldering iron comes in, and there are all kinds, depending on the work to be done and on the pocketbook.

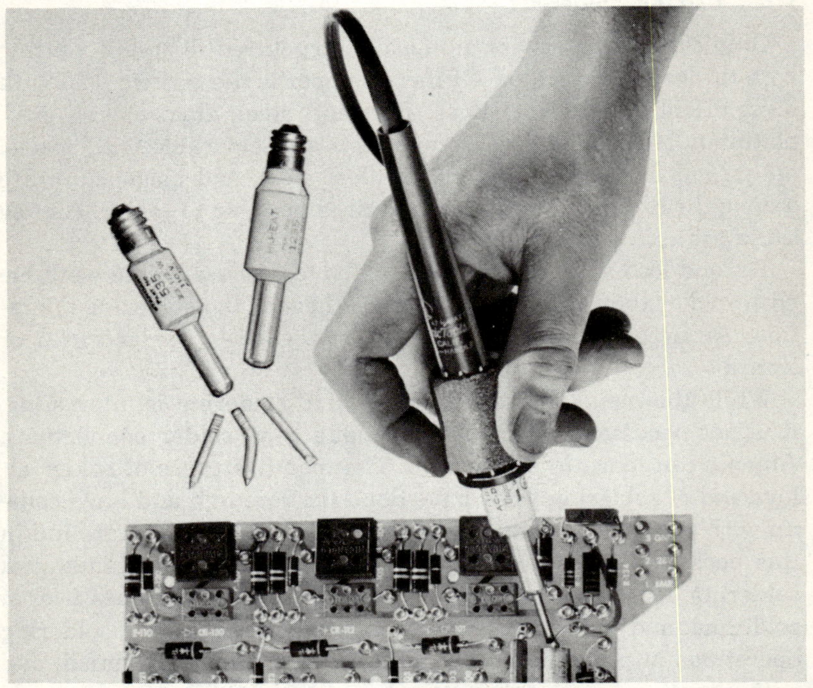

Courtesy Ungar Division of Eldon Industries, Inc.

Fig. 1-6. Ungar interchangeable heating elements and tips.

REPRESENTATIVE SOLDERING IRONS

A few representative soldering irons will now be considered, though they will be covered in greater detail in later chapters.

Standard-size soldering irons with a rating of 75 watts or higher have not changed much over the years. Electrical circuits using point-to-point wiring from tie points to device terminals are more easily soldered with the high heat of these irons.

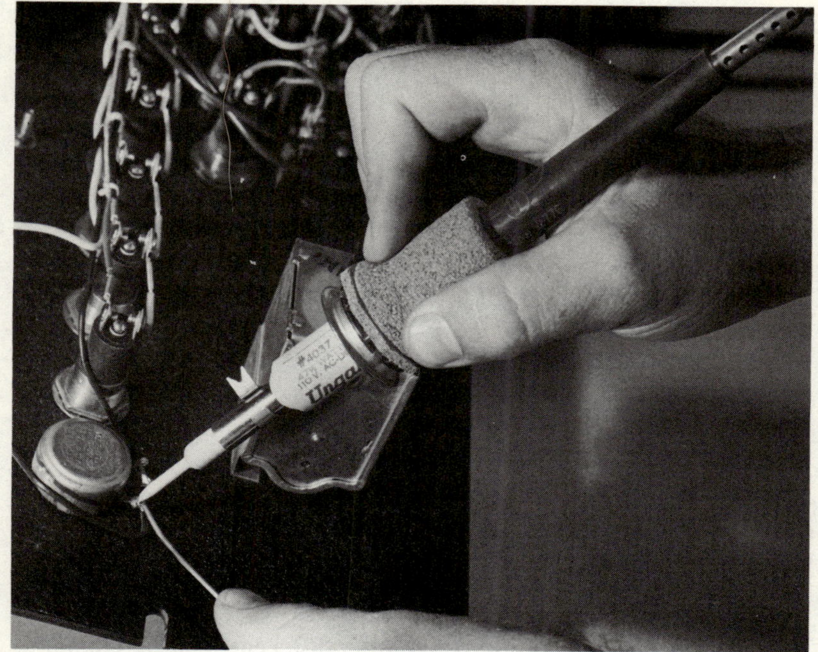

Courtesy Ungar Division of Eldon Industries, Inc.

Fig. 1-7. Standard Ungar No. 776 handle.

Today's electronic circuits that use semiconductor devices, in most cases mounted onto printed-circuit boards, require soldering irons of smaller point size and lower heat ratings.

Interchangeable heating elements, tip sizes, and tip configurations have become popular. The Wall Brand iron shown in Fig.

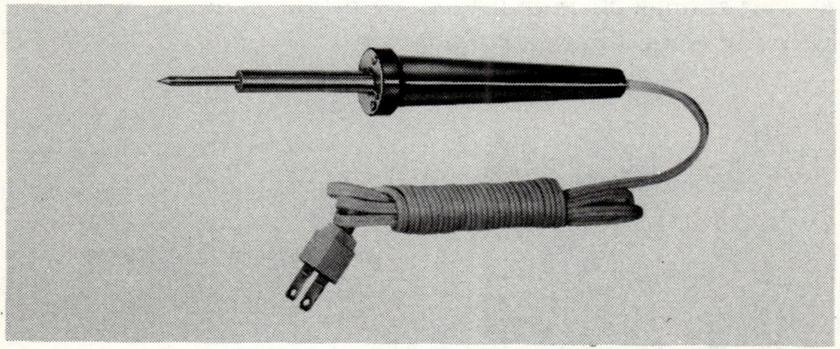

Courtesy Cooper Industries

Fig. 1-8. The Weller 25-watt soldering iron is typical of the new "pencil" iron. It is small, lightweight, and held in the hand like a pencil.

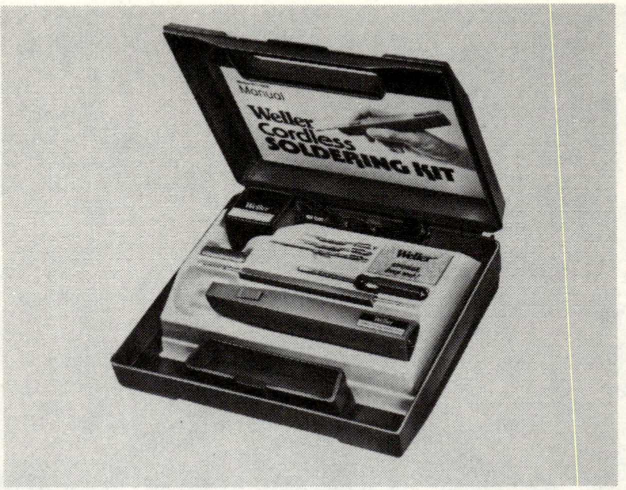

Courtesy Cooper Industries

Fig. 1-9. Cordless soldering irons are now made by many soldering iron manufacturers. The Weller kit shown here is complete with an iron having built-in nickel-cadmium batteries, a charger, three tips, and other accessories.

1-5 uses a single handle into which nine different heating elements can be plugged. The handle is made of blue polycarbonate, which stays cool. Elements with ratings from only 11 watts up to 75 watts can be used in the handle. The four lower-wattage elements have ⅛-inch diameter tips, and the five higher-wattage elements have ¼-inch diameter tips.

The interchangeable heating elements, some of which also accommodate interchangeable tips, are a special feature of the Ungar line. Fig. 1-6 shows heating elements with 23½-, 37½-, and 47½-watt ratings. Each is tapped for a ⅛-inch screw, which permits use of eight different styles of tips.

Fig. 1-10. Wen Model 450 soldering gun.

Courtesy Wen Products, Inc.

Fig. 1-7 shows the standard Ungar No. 776 handle with one of three interchangeable heating elements with ¼-inch male threads. These threads take four styles of tips with shank diameters from 0.322 inch down to 0.125 inch.

Weller is a well-known name in soldering irons. The iron shown in Fig. 1-8 is their standard 25-watt iron, and is only one of many sizes and styles. A new addition to their line, as well as to other brands, is a cordless soldering iron. The iron operates from low-voltage nickel-cadmium (Nicad) batteries built into the handle.

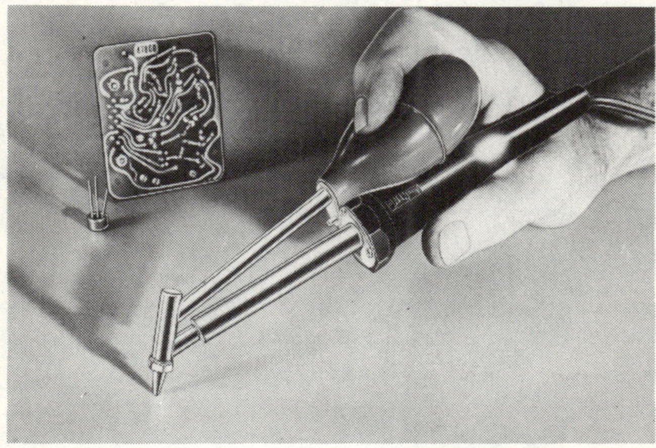

Courtesy Enterprise Development Corp.

Fig. 1-11. An Endeco desoldering iron.

When not in use, it is plugged into a charger connected to the 120-volt ac line. Fig. 1-9 shows a cordless soldering iron kit that includes an iron with built-in nickel-cadmium batteries, charger, three tips, and other accessories.

Both Weller and Wen make an "instant-heat" iron, which is shaped like a gun and is usually called a soldering gun (Fig. 1-10). The body of the gun contains a transformer that steps 120 volts down to a low-voltage, high-current supply for the tip. This low-voltage high current flowing through the tip will heat it in 3 to 10 seconds.

Many irons also include a built-in desoldering function. Fig. 1-11 shows a typical desoldering iron. The desoldering iron is used to remove solder from a solder connection—especially from a printed-circuit board—so that a defective part can be removed without damaging the solder-connection area. To use the desoldering iron, simply place the iron, with the vacuum bulb depressed, over the solder connection. When the solder melts, release the vacuum bulb, and the solder will be drawn up and away from the connection.

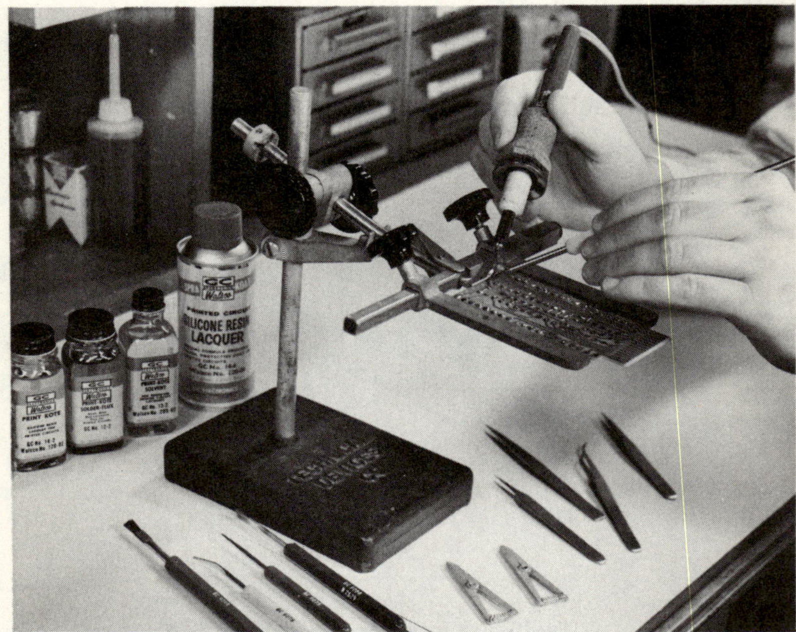

Courtesy GC Electronics Co.

Fig. 1-12. Soldering aids.

Fig. 1-12 shows special soldering aids that make repairing easier. From right to left are: tweezers for grasping fine wires or holding small parts, heat sinks for semiconductor leads, special picks for desoldering, and a brush with stainless-steel bristles for cleaning away excess solder.

2

The Solder Alloy

Solder is an alloy of two or more metals used in various combinations to join base metals together by the application of heat. In all cases the applied heat melts the solder, but not the base metal. Thus, soldering is a workable method of joining metals easily.

SOFT SOLDER

While there are many "solders" or solder alloys, this chapter is concerned only with the popular "soft solders" used in electrical connections. Soft solder is essentially an alloy of the following two metals:

Metal	*Chemical Symbol*	*Latin Name*
Tin	Sn	Stannum
Lead	Pb	Plumbum

Like water and alcohol, liquid tin and lead can be mixed together in a solution. The solution can be any ratio of tin and lead. Such a mixture is called an *alloy*. An alloy retains most of the characteristics of the original metals, although a few changes take place. There may be a change in melting point, a change in hardness, and, in the case of solder, a difference in wetting ability.

Tin is not attacked by air or water. This property of tin permits it to be used frequently as a coating on copper wire leads for protection against corrosion—thus the "tinned" leads on resistors and capacitors.

Lead is soft and dense, but its surface quickly corrodes. The corrosion at the surface, however, acts as a barrier to further corrosion below the surface, and this action becomes one of the good qualities of lead as a coating.

Tin reacts and alloys with the joining metals easily. However, its melting temperature is too high for it to be used alone as a solder, and it is too costly. Lead reduces the melting temperature and, when alloyed with tin, adds to the overall strength, although it moderates the action of alloying to the base metal.

The proportions of tin and lead in solder are expressed as a ratio. The first two digits represent the percent of tin, and the last two digits represent the percent of lead. For example a 60/40 solder means it is an alloy of 60 percent tin and 40 percent lead. It may contain traces of other metals, but this will not be shown by the figure.

Under the ASTM (American Society for Testing Materials) specifications, a different grading or numbering system is used. Only the percent of tin is given, not the percent of lead. If a letter suffix is shown, it will be the amount of antimony, the most common additive of solder. The suffix A denotes 0.12 percent or 0.25 percent of antimony; the suffix B, 0.50 percent; and the suffix C, 2.0 percent. Under ASTM specifications, a 60/40 solder with 0.5 percent of antimony added would be designated 60B. Obviously, the number 60 subtracted from 100 leaves 40 percent for the lead content.

Here are some of the tin/lead ratios and their properties:

70/30	Good for pretinning; expensive
63/37	Eutectic solder; no pasty state; best solder
60/40	Good general-purpose solder
50/50	Lower-cost general-purpose solder
40/60	Lowest-cost common solder

TIN-LEAD TEMPERATURE GRAPH

The graph in Fig. 2-1 shows the conditions of solders at various temperatures. Note particularly the area called *pasty range*. Except at one precise ratio of tin and lead, alloys of tin and lead pass through a stage of pastiness when cooling from a liquid state to a solid. In this state the alloy is neither a liquid nor a solid, but has a mushy consistency between a liquid and a solid. Pure tin and pure lead change from liquid to solid without the pasty stage. A ratio of 63 percent tin and 37 percent lead, called the *eutectic ratio,* also changes from liquid to solid without passing through the pasty stage.

The horizontal line in the chart called *solidus* at 361°F is the temperature below which all tin-lead solders will solidify as they cool. The *liquidus* lines represent the bottom temperature for melting the solder alloy into a liquid. While all common solders solidify at the same temperature, the various alloys require more heat to melt, depending on their ratio of tin to lead. Only the 63/37 eutectic ratio has the lowest temperature for melting to a liquid.

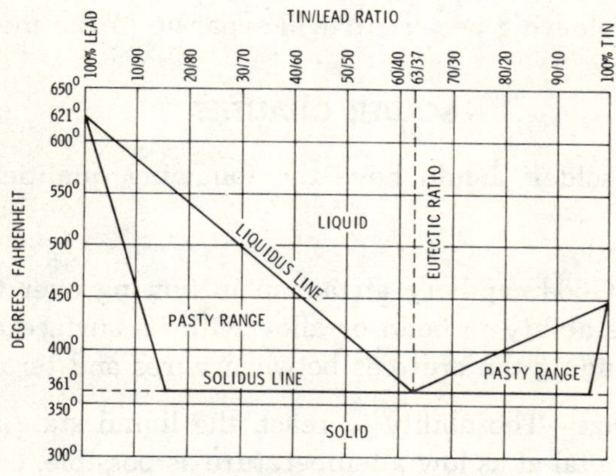

Fig. 2-1. Tin-lead temperature graph.

The popular 60/40 solder used for quality general-purpose electrical soldering is very nearly like eutectic-ratio solder, having a low liquidus temperature. Note the higher liquidus temperature for the less expensive and lesser-quality 40/60 solder. It takes a hotter soldering iron to melt this solder.

For most soldering purposes, the eutectic ratio of solder is best. If, for example, the wires of a connection are disturbed while the solder is cooling and in the pasty stage, the result is a weak connection. This characteristic could be important to beginners, as well as to industry. Because of the higher tin content, eutectic solders are priced higher than other types. Eutectic solder has the lowest melting point of tin-lead solders (361°F), an important consideration in soldering.

SOLDER ADDITIVES

Most solders include a small amount of one or more other metals that give them special characteristics. For example, adding from 0.25 to 2 percent of antimony hardens the solder and reduces its tendency to "creep" in the cold state. The antimony does, however, slightly reduce the solder's ability to wet or flow into the base metal. Adding copper reduces the pitting of copper tips on soldering irons. Adding a little silver or gold retards the reduction of silver or gold from connections plated with these metals. Many other metals may be added in small amounts for special purposes. Small amounts of cadmium are sometimes added to lower the soldering temperature. The right amounts of tin, lead, and cadmium can be eutectic at 265°F. Cadmium cannot be used in solder pots or dip soldering because it will separate in the molten state.

SOLDER QUALITIES

Ideally a solder should have the following qualities for best performance:

Wetting—Good capillary attraction in flowing over the base metal. The ability to bond or alloy with its surface at every point, including the crevices between wires and terminals.

Temperature—The ability to reach the liquid state and wet the base metal at as low a temperature as possible, consistent with its other qualities. Minimum temperature means least possibility of damage to the terminals or components being soldered.

Strength—A degree of hardness as high as possible, consistent with other qualities. Freedom from brittleness, a minimum of coefficient of expansion compared with the base metal, and low creepage.

No solder is perfect in all the above requirements, but modern alloys come close to the standards. The better known solder producers can be depended on to supply the best that the modern science of metallurgy makes possible. The use of solder connections in critical aerospace and military electronic equipment attests to this assertion.

It is important to keep in mind the meaning of the word *wetting*. Wetting is the action between the molten solder and the base metal being soldered. It is similar to the capillary attraction that draws the water up into a towel. Just as this capillary attraction varies between water and various types of cloth, so does the ability of solder to wet certain metals. Certain solders wet better than others.

Lead is poor in wetting ability. Tin is good. Combinations of tin and lead provide varying qualities of good wetting, economy of product, ability to wet at low temperatures, and hardness.

Wetting must not be confused with tinning. Tinning means a *precoating* of the base metal with solder, whether it is pure tin or an alloy of tin and lead. Wetting is the flow of solder over the surface of the base metal, and the bond at the interface of the two. Atoms of the solder make a chemical bond and alloy with the surface atoms of the base metal.

THE FORMS OF SOLDER

Rosin-core solders are available in small packs for the home user. Some of these solders come in dispenser-type containers. The most popular size for the serviceman's bench is the 1-pound spool. Solder can be purchased in 1-, 5-, and 20-pound spools for industrial use.

The system for sizes varies with the processors, some using wire-gauge figures, others using diameter in inches. The most popular wire sizes and alloy ratios available from the stock of solder producers are (any ratio in any size listed):

Wire Size or Diameter		*Ratios Alloy*
1/16″ (.062″)	In any of the ratios	63/37
1/32″ (.031″)		60/40
No. 16 (.064″)		50/50
No. 18 (.048″)		40/60
No. 20 (.036″)		

The wire size to select depends on the size of the soldering connection to be made. The smaller sizes permit better control of the amount of solder being fed to small connections, such as those on printed-circuit boards. Smaller-size wires melt faster because less of the heat is carried away by the rest of the solder. Better control is had with larger wire sizes on large connections such as terminals and solder lugs, and these sizes are more economical.

Most solders are purchased with a continuous inner core of flux. Some are multicored, with as many as five cores of flux. Solder is also available without the inner flux core, for use with separate liquid flux if preferred. The flux cores also are different, depending on the activity desired. The subject of fluxes needed for soldering is covered in the next chapter.

In the electronics industry, the hand soldering of electrical connections is done almost exclusively with flux-cored solder. The amount of flux can be carefully metered out this way. There are times when a separate flux is desirable in hand wiring, but this is mostly confined to the soldering of printed-circuit boards by the dip or wave process.

The figures in Table 2-1 show the number of feet in a pound of solder wire in various sizes. These figures may vary somewhat with different brands.

ALUMINUM SOLDERING

In electrical and electronic work, it is sometimes necessary to solder aluminum to copper, or aluminum to aluminum. An example of the latter is the installation of an aluminum baffle in an

Table 2-1. The Length of One Pound of Rosin-Core Solder Wire

Wire Gauge Size	Inches Dia	Solder Tin/Lead Ratio		
		60/40	50/50	40/60
18	.048	182 ft	174 ft	167 ft
20	.036	324 ft	319 ft	299 ft
22	.028	536 ft	529 ft	493 ft
24	.022	865 ft		
26	.018	1290 ft		
28	.015	1910 ft		
30	.012	2730 ft		
32	.011	3590 ft		
34	.009	4950 ft		

Courtesy Multicore Sales Corp.

aluminum box to shield one part of a circuit from another. This can be done with special-purpose solders, but it requires higher heat than electrical soldering does, but much lower heat than brazing or welding.

Kester makes a special AL Flux-Core aluminum solder. It is recommended only for use on pure aluminum such as grades 2S, 3S, and 4S. Ersin Multicore calls their solder Alu-Sol. It is not recommended for use on aluminum with more than 0.5% impurity.

3

Fluxes

If a freshly brightened piece of copper were maintained in an inert atmosphere, there would be no problem in making solder connections to it without the aid of flux. Because of the inert atmosphere, the surface of the copper would remain pure, and heated solder would wet it with ease. Under normal working conditions, however, the oxygen in the atmosphere attacks copper, and forms a thin layer of copper oxide. When copper is exposed to the air long enough, the oxide appears as a green tarnish. But even if the copper is exposed only a few seconds, there is enough oxide formed to prevent soldering to the copper. And that's where the use of a *flux* comes in.

A soldering flux acts on the oxides that form on the surface of many metals. The thin layer of tarnish becomes soluble in the flux and evaporates when the flux is heated to its boiling point. The action of the flux is shown in the sketch in Fig. 3-1. In this example a hot soldering iron, well-tinned with solder, is drawn across a metal surface. As the iron is moved across the surface of the metal, its heat melts the solder and its inner core of flux, causing the flux to boil at the iron. The boiling flux reacts with the tarnished surface and floats the tarnish away to be evaporated with the flux. At that point no new air can reach the metal, so it stays untarnished and the molten solder alloys with the heated surface of the base metal.

In a normal solder connection, the soldering iron is not drawn across the surface of the connection, but held against the wire connection. As flux-cored solder is held against the heated connection, the flux melts out of the core, and the liquid flux spreads across the surfaces of the connection. As the iron is held against the terminal, the flux heats up still more, reacts with the tarnish of the base metals, floats it away, then boils out into the air. As the iron heats the connection still further, the solder melts and flows into the crevices of the connection.

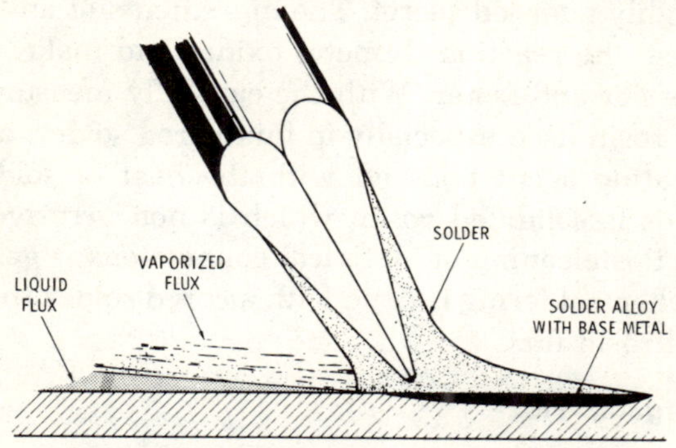

Fig. 3-1. How flux acts on surface oxides.

ORGANIC AND INORGANIC FLUXES

Fluxes used in soldering are divided into two general categories, organic and inorganic. Certain organic fluxes are used in electrical soldering, and inorganic fluxes are used in sheet metal work and plumbing. Most organic fluxes are noncorrosive. Inorganic fluxes, though much more active than organic types, are corrosive to various degrees.

Organic Rosin or Resin

The words *rosin* and *resin* are used interchangeably, and both words are used by processors of soldering fluxes. Specifically, *resin* is a general term for an organic solid or liquid, insoluble in water. *Rosin* is the residue from the distillation of the turpentine of white pine. You may have seen it used as a solid and rubbed on the bows of violins, or crushed into a dust and spread on the canvas of a fight ring to increase the grip of shoes on the canvas. When diluted in a small amount of turpentine, it is "plastic rosin." When diluted in alcohol, it is a liquid flux for soldering.

Rosin is chemically an abietic acid. Although not corrosive to metals, it does react with the oxides of metals. It is the original and still most used basic flux for electrical soldering. It is sometimes referred to as *water-white gum rosin.*

Rosin, by itself, is not a very active flux. That is, it is not conducive to easy wetting by the solder. Most flux processors add a form of amine hydrochloride or other ingredients to it to increase its activity and make soldering easier. The exact activator used in the flux is proprietary to the processor, and its composition is usually a highly guarded secret. The ingredients of an "activated" rosin increase the reaction to metal oxides and make wetting by the solder easier and faster. With the carefully measured amount of activated rosin used especially in flux-cored solder, any residue of the activating agent boils off with the heat of soldering, and what remains is solidified rosin, which is noncorrosive and nonconductive. Postcleaning of electrical connections is generally not necessary when soldering is done with a cored solder that contains an activated rosin flux.

Nonrosin Organic Flux

There are some fluxes that do not use a rosin for a base. Amine or hydrochloride bases, mixed with other ingredients, make an active flux. Although their soldering activity is generally greater than that of rosin-base fluxes, they are slightly corrosive to the base metal. Most of the residue is boiled away at the time of soldering, but postcleaning of all solder connections is recommended.

Both types of organic fluxes are sensitive to excessive heat, and because of this, care is required in making good solder connections.

Inorganic Fluxes

Inorganic fluxes are forms of acids or salts, which, while highly active in the removal of tarnish from metal, are also corrosive to various degrees, and are not used for electrical soldering (with one exception, discussed in the next paragraph). An example of an active, inorganic flux is zinc chloride, which is made from zinc and hydrochloric acid, with some additives. The activity of such fluxes is so high that heavy pieces of metal can be soldered. They are the fluxes used by sheet metal workers. Since they are highly corrosive to metals, and since they do conduct electricity, inorganic fluxes should not be used to solder electrical connections, unless a thorough washing can be applied to remove every trace of the flux afterwards. Although these fluxes are usually referred to as acids, they are really salts. However, the residue left after soldering is hygroscopic and will pick up moisture from the air. The flux then becomes acid, and thus corrosive.

If electrical soldering *must* be done at a high temperature, it may be necessary to use one of the inorganic fluxes. The rosin-base fluxes char at high temperatures and lose their ability to react with tarnish of the base metal. The inorganic or acid fluxes do not char and are not affected by high heat, but it is absolutely necessary to neutralize the corrosive action, or wash the flux away with detergents or other means. Outside of this one exception, acid fluxes must be avoided in making electrical solder connections.

CLEANLINESS

Fluxes will remove the thin tarnish resulting from the exposure of a base metal to air if the metal has been exposed for only a short period. Fluxes will not remove greasy or oily surfaces, or the scales of other corrosive atmospheres, or the results of the oils and acids of human perspiration. Before soldering, the work must be clean. If the work has been in storage for a long time, it may require cleaning with an abrasive. If it has been exposed to machine operations that resulted in an oily atmosphere, the work must be cleaned with a degreaser or washed with a detergent. It is preferable that component connecting leads not be handled by bare fingers.

FLUX CORES

By far the majority of soldering is done by hand with soldering irons and a wire solder having a flux core. Today, all rosin-base fluxes in the core are fairly active and safe to use on electrical connections. Different producers may use different additives to make the flux more active, but all fluxes do a good job of easy wetting by the solder. The cores themselves differ with the various

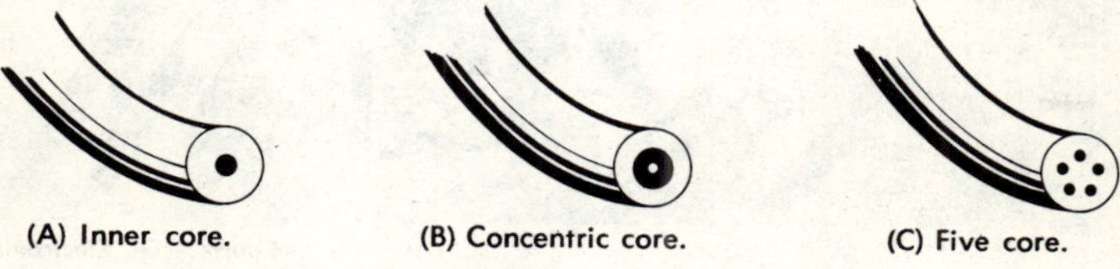

(A) Inner core. (B) Concentric core. (C) Five core.

Fig. 3-2. Types of flux-core solders.

producers. A cross section of the solder wire will look like some of the drawings in Fig. 3-2. The dark areas shown in the sketch are the flux; the light areas the solder. The most common core is in the center, as shown in Fig. 3-2A. The flux occupies the center hollow of the solder, which is formed like a pipe. Some producers claim superior results with a concentric core like that in Fig. 3-2B. They say that a better mix is obtained with solder on each side of the flux. Others favor the three-core and five-core systems, saying that there may be voids in the solder core, and a multiple core reduces the chance of a void affecting the solderability of the solder.

NONCORE SOLDERING

With the increasing use of printed-circuit boards in electronic circuits, industry has developed more and more ways to rapidly solder entire systems. *Dip* and *wave soldering* involve dipping the entire board into a molten bath of solder. The flux, then, must be applied in advance by brushing on a liquid flux or by dipping the board into a flux bath. Mass fluxing calls for careful engineering

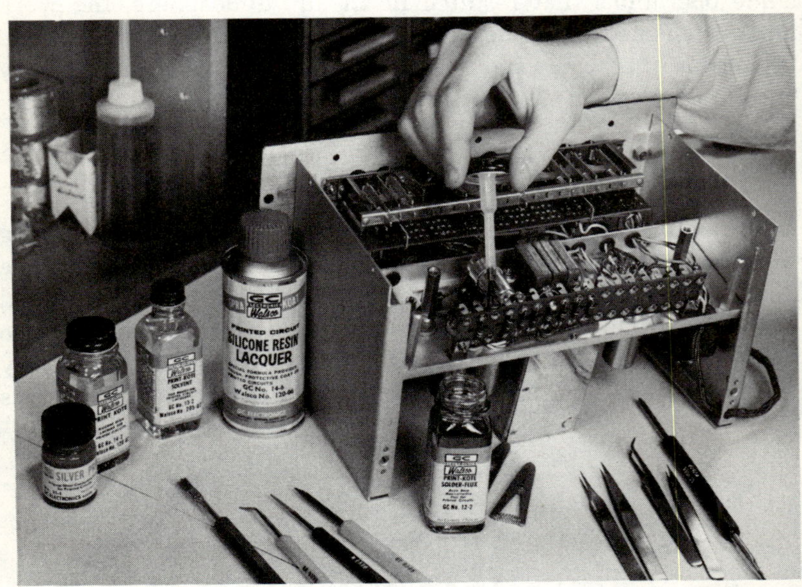

Courtesy GC Electronics Co.

Fig. 3-3. Using a liquid flux.

of the technique so that the right amount of flux is applied. In dip fluxing, no postcleaning is needed with the use of pure rosin. It may be necessary to do a cleanup job after dipping the board into an activated flux. Industrial soldering of this type calls for working closely with the supplier of fluxes and solders, as well as with the suppliers of printed-circuit boards.

Resoldering old terminals is sometimes easier if a liquid flux is applied to the terminal first (Fig. 3-3). Excess flux is easily removed by wiping with a cloth dampened with alcohol after the connection has cooled.

4

Soldering Irons

Early soldering irons were heavy pieces of copper used to transfer heat from a flame to the metal to be soldered. These early soldering irons had a handle, and their far end was ground to a pyramid shape for soldering into corners. But that was before the electronics era.

The growth of electronics, the miniaturization of components, and the use of printed-circuit boards made improved types of soldering irons necessary. Gasoline-torch-heated irons gave way to internal electrically heated irons. Entire new types of soldering tools and superior techniques were developed. One such type is shown in Fig. 4-1. Weller irons range in wattage from the 25-watt unit shown in Fig. 4-1A to 175-watt units. Wall-Lenk Manufacturing Company produces a comparable line of soldering irons ranging up to 200 watts (Fig. 4-1B).

Another type of soldering iron is the quick-heat gun. In this type, the heat is generated at the tip. Similar to the quick-heat gun is the pistol-grip semiquick iron. Both of these are on only while the trigger switch is held on. These will be discussed in more detail in Chapter 5.

The most remarkable development engendered by the electronics era has been the miniature iron with its tiny tip for making printed-circuit connections. Fig. 4-2 shows three irons made by American Electrical Heating Company. These are designed for soldering small connections, using tips from ⅛-inch diameter to ¼-inch diameter. American Beauty is probably the oldest manu-

(A) A Weller 25-watt iron, packaged with interchangeable tips and other soldering accessories.

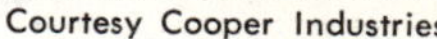
Courtesy Cooper Industries

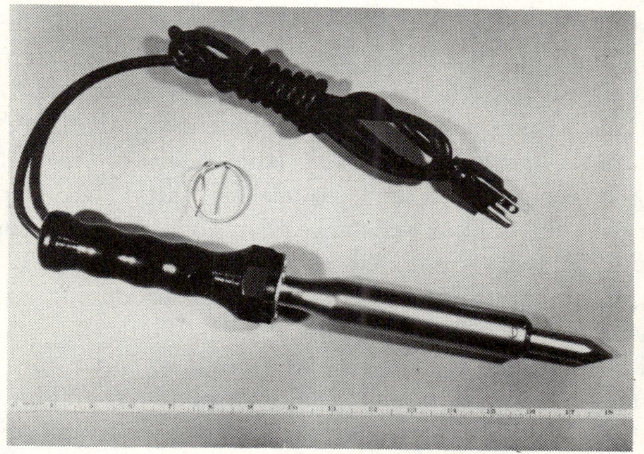

Courtesy Wall-Lenk Manufacturing Co.

(B) A Wall 200-watt soldering iron.

Fig. 4-1. A small soldering iron and a heavy-duty soldering iron.

facturer of soldering irons. Their irons are specially constructed to permit replacement of any part of an iron, when necessary.

For years plumbers have used gasoline blowtorches for their soldering. Now for electrical soldering we have butane torches with better regulation of the flame. This is the subject of a separate chapter.

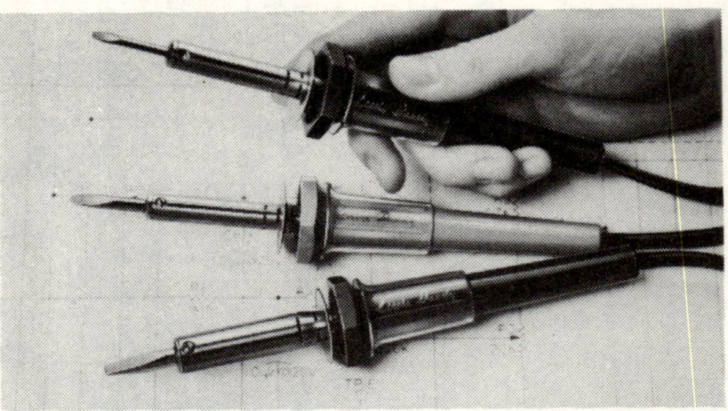

Courtesy American Electrical Heater Co.

Fig. 4-2. American Beauty "Little Dandy" pencil-type soldering iron. The heating element is available in ratings from 20 to 40 watts, and the tip comes in 1/8- to 1/4-inch diameter sizes.

SOLDERING-IRON CONSTRUCTION

In the basic soldering iron, heat is conducted from inside the barrel to the tip. A coil of nichrome wire develops the heat. The size and length of the wire determine the amount of current that will pass through it when 120 volts is connected across the ends. The wattage, which is turned 100 percent into heat, is a function of the formula in which $E^2/R = W$ (the square of the voltage divided by the resistance in ohms equals the watts). Nichrome (and similar wire) has higher resistance than copper for the same size and length, so it does not take as much of it to make a coil. In addition, it is not subject to oxidation when producing high heat, as copper is.

A coil of nichrome wire is wound around an inner core of copper, although the wire is electrically insulated from the copper. The heat of the coil is carried by the copper core to the tip, which fits snugly into the core. Copper is a very good conductor of heat. A protective shell, a handle, perhaps heat radials, and a line cord and plug complete the basic soldering iron (Fig. 4-3).

Of course, the construction of a soldering iron is not as simple as it appears here. The requirements of long life, a cool handle, efficient transfer of heat to the tip, and good heat recovery call for special construction design. This applies whether the iron is large or miniature (Fig. 4-4). Most good-quality large irons have a 100-watt or better rating (Fig. 4-5). Some of the less expensive household irons in the larger-than-miniature sizes are rated at around 50 watts. Most of these are low-priced and intended only for occasional soldering around the home.

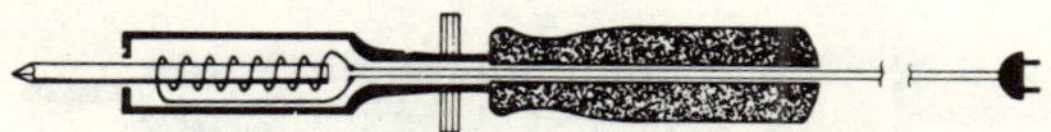

Fig. 4-3. Cutaway view of soldering iron.

PENCIL-TYPE IRONS

Most printed-circuit boards have very small connections, and the larger irons are clumsy for soldering to such small connections. Furthermore, unless expertly used, the irons may be too hot for the safety of the copper laminate on the board and for

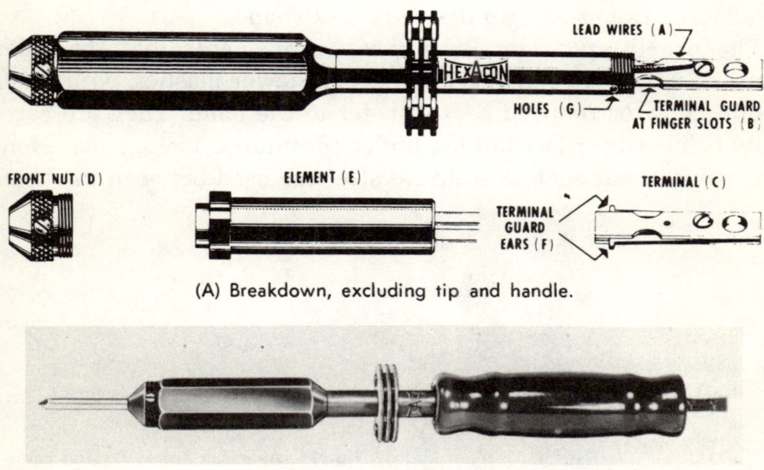

(A) Breakdown, excluding tip and handle.

(B) Assembled soldering iron.

Courtesy Hexacon Electric Co.

Fig. 4-4. Construction of soldering iron.

semiconductors wired to the boards. This difficulty has resulted in the development of miniature irons that have small tips for the small connections. These irons are lighter in weight, easier to hold, and lower in power than the larger irons. They are referred to as *pencil irons* because they are held like a pencil. Fig. 4-6 shows an

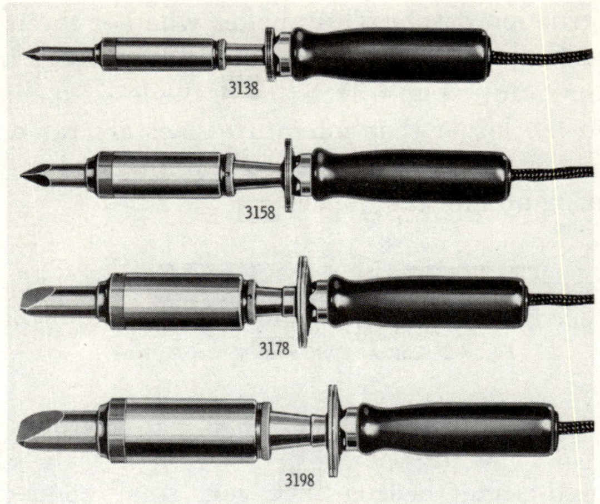

Courtesy American Electrical Heater Co.

Fig. 4-5. Four models of American Beauty line, having ratings from 100 to 550 watts.

American Beauty miniature iron, available in 10 different models of wattage rating and tip diameter and shape.

The pencil-type irons have many refinements that the older irons do not. Their handles are made of newer plastics, with inner cooling fins for reduced heat transfer to the hand. They are carefully balanced for less fatigue under continuous factory use. Noncorrosive metals such as stainless steel are used between the heat-

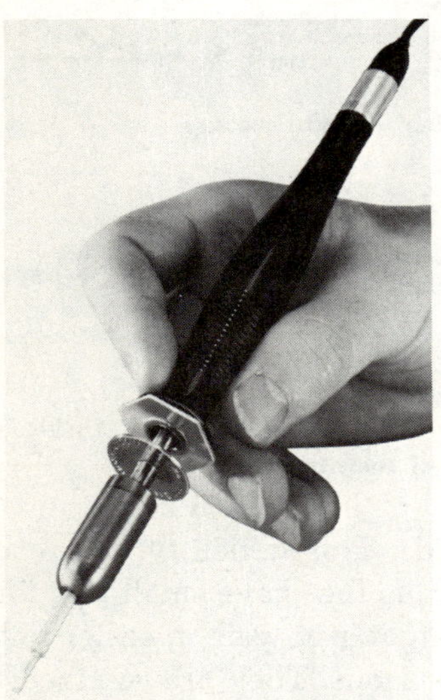

Courtesy American Electrical Heater Co.

Fig. 4-6. American Beauty B-series miniature iron.

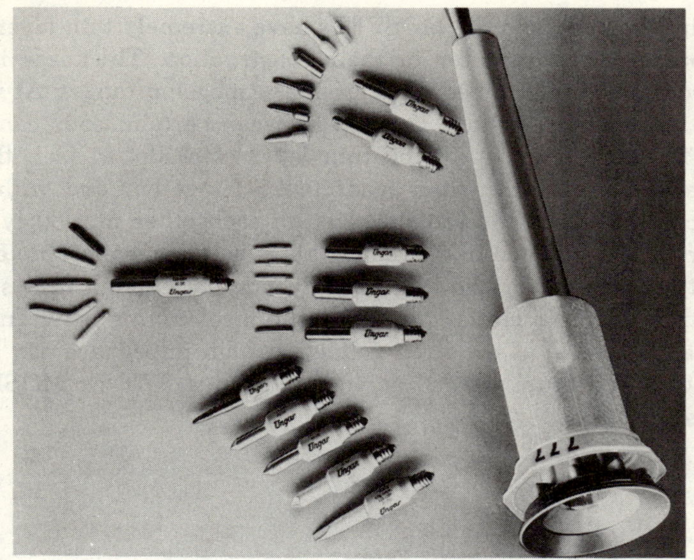

Courtesy Ungar Division of Eldon Industries, Inc.

Fig. 4-7. Ungar handle accommodates many heating elements which accommodate a variety of tips.

ing sections and the tips for easier tip replacement. Probably the most popular line of pencil irons for general-purpose use is that of Ungar, in which the handles have a candelabra base-screw connection, allowing interchangeability of heating elements as well as tips (Fig. 4-7).

Many irons, especially those for industrial use, are now supplied with a three-conductor cord and plug. One of the leads in the cord grounds the exposed metal of the iron, to prevent electrical-shock hazard to the user. Highly improbable, but always possible, is the shorting of part of the inner heating coil to the outer metal shell, which could result in a voltage potential between the iron and a grounded workbench. With the outer shell grounded through the third conductor of the line cord, such a short would result in the blowing of a fuse, but there would be no shock hazard to the person using the iron.

TRANSFORMER-ISOLATED IRONS

Semiconductors such as FETs and MOSFETs can be easily ruined by a soldering iron. The FET transistors have very high input impedances, and the MOSFETs have extremely thin layers of an oxide as an insulator in their construction. Thickness is in microns, and input impedances are in the megohm ranges. At such high impedances, voltage buildup between certain leads can result in the puncturing of the thin layer of oxide. In fact, these transistors come with their leads twisted together and with instructions to keep the leads shorted to each other until they are inserted into position in a circuit. This is because even static electricity picked up by the human body may be enough to destroy the transistor. No matter how good the insulation is between the coil and the rest of the iron, voltage at high impedance can leak out to an external device, enough to ruin a FET or MOSFET transistor.

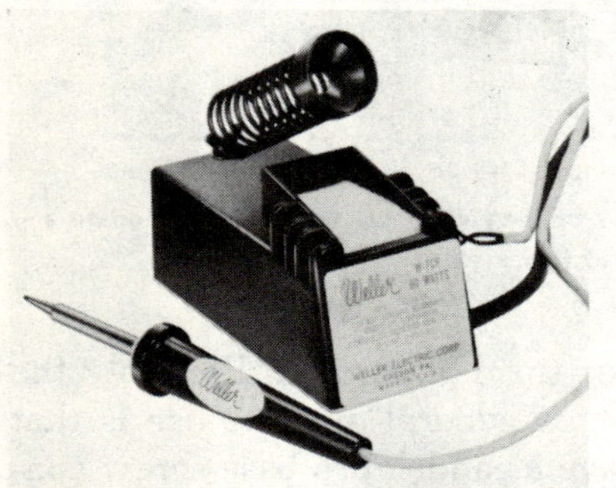

Courtesy Cooper Industries

Fig. 4-8. In the Weller Model WTCP, a transformer supplies 24 volts (ac) to the pencil iron.

The need to protect these high-impedance transistors has resulted in the development of soldering irons that work at lower voltages, and from an isolating transformer (Fig. 4-8). Voltages of 12 and 24 volts are commonly used. Not only is the lower voltage a lower source of leakage voltage, but the transformer used to drop the voltage provides additional isolation from the 120-volt ac line. This additional isolation reduces the possibility that ac line surges and resulting peak pulses will reach the transistor leads while they are being soldered into place.

IRON COSTS

The cost of an iron is a function of its quality. It is true that the higher the wattage, the higher the cost, because of the greater amount of material used; the barrel with the heating element is larger, as are the copper reserve core and the tip. But more than size, quality of construction determines cost.

Manufacturers assembling electronic circuits need an iron that will provide long life under constant use. The efficient transfer of heat from the heating element to the tip is important. The iron must stand up under mechanical abuse. It must be light in weight, fit the hand comfortably, and stay cool. Handles are made of polypropylene or other plastic that does not conduct heat and that stands up to handling. This is especially true for the miniature irons, even though cork is a perfectly adequate handle for occasional home use. The ability to interchange tips of various shapes also affects cost.

For the kit builder or occasional electronics home experimenter, good results are possible even with inexpensive irons having only a single tip. A miniature iron of moderate power between 25 and 40 watts is good for most electronic work, whether it is used on printed-circuit boards or for standard electrical connections on terminals. Such an iron with about a ⅛-inch-diameter tip shank and a pyramid-shaped point, or modified pyramid, will be satisfactory for almost any job, except making a solder connection directly to a metal chassis. The tip can be copper, which will require only an occasional redressing and retinning.

CONSTRUCTION DETAILS

The handle of a soldering iron must be more than just a device for holding the iron while working with it. It must be comfortable, and it must stay cool. Many irons have metal fins between the handle and hot part of the iron. These fins dissipate the heat from the shank that connects the iron to the handle. Some handles are fluted on the inside to allow a flow of air under the outer handle material for cooling.

Constant-duty irons have heating elements that are replaceable. Although the heating elements will often last forever, they are subject to some deterioration with constant use. Outer shells can come apart, and the removable heating element can be disconnected from the line cord by means of screw terminals. In some miniature lines, the entire heating element can be unscrewed from the handle, without disassembly of the iron.

Tips are interchangeable in all but the cheapest of irons. In the larger-sized irons, the tips have either straight shanks or screw threads at one end. Straight shank tips can be inserted into the hollow inner core. The sleeve fits closely to provide good heat transfer. Stainless steel is most often used for the inner sleeve lining to prevent freezing (Fig. 4-9). The tips are held in place by setscrews or knurled nuts and split threads, for straight shank tips. Screw tips can be screwed onto the end of an inner copper core, but the threads are plated to prevent freezing.

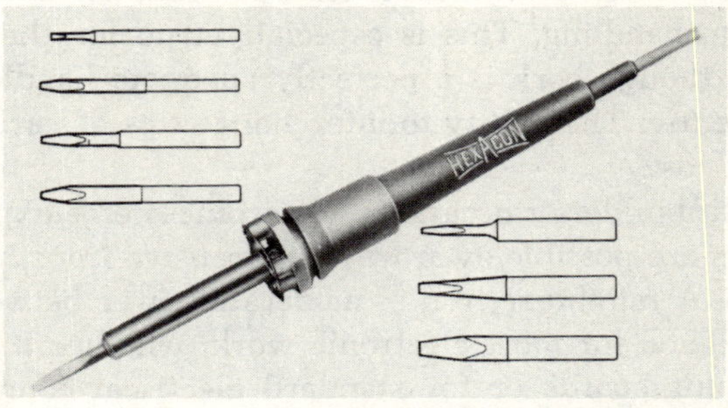

Courtesy Hexacon Electric Co.

Fig. 4-9. Hexacon's "Slim Junior" takes a variety of straight-shank interchangeable tips.

The number and variety of interchangeable tips are greatest for pencil-type irons. They are usually threaded onto the inner core.

HEAT CONTROL

A certain amount of heat control is natural to all soldering irons. All metals expand with heat. When metal wire is heated, its resistance increases, resulting in a decrease in wattage and heat. With the nichrome wire used in most irons, the coefficient of expansion is small, and so is the wattage regulation. Copper has a higher coefficient of expansion and could provide good regulation, but copper oxidizes in air, and oxidizes very rapidly when hot.

Another form of natural regulation is the dissipation of the heat from the metal parts of the iron to the surrounding air. As the iron heats up, even if not used to make solder connections, a point of equilibrium is reached when the heat dissipated into the air is equivalent to the heat generated.

Delivering the right amount of heat to a connection is important for making a good connection. Automatic heat regulation is incorporated into some irons for this. One method is a built-in bimetallic strip, which cuts off current to the heating element when the temperature reaches a certain value, and reestablishes contact when the iron cools a few degrees. Another method uses a magnetic core (Fig. 4-10). When a mating core is cool, the magnetic core is attracted to it and closes contacts to the heating element. As the mating core warms up, it loses its ability to attract a magnet and lets go of the magnetic piece, and the contacts open.

In a regulated iron, the right heat is available at the tip at all times. As heat is drawn from the tip, the contacts will close and may even stay closed with the iron in continuous use. When the iron is laid aside when not in use, the lower heat will prolong the life of the tip and, of course, conserve on power drawn from the ac line. This is important only in factory production work where many irons are in use throughout the working day.

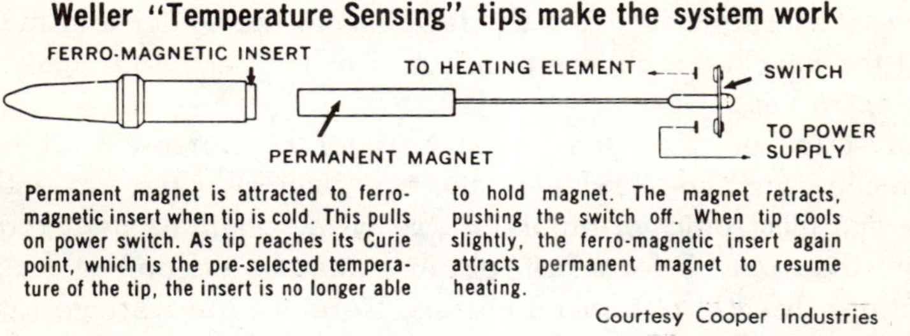

Fig. 4-10. Description of Weller magnetic "temperature-sensing" system, taken from their literature.

Certain very fine soldering (printed-circuit boards, etc.) calls for regulated and adjustable tip temperatures. These features are built into deluxe-type soldering-iron stands which are available from several manufacturers. Some soldering-iron stands even include solid-state regulating circuits. See Chapter 8 for some examples of these.

Heat Requirements

While the wattage rating of an iron is a relative indication of heat that can be available, the actual heat reaching the tip may vary, depending on the heat-transfer efficiency of the iron, the smallness of the tip diameter, the distance from the source of heat, and the work being done; that is, the rate at which heat is being drawn out of the tip while soldering. This last point is a function of the size of the connection being made, and it is the all-important factor in determining what iron to use.

Solder usually becomes liquid at temperatures between 361° and 370°F, depending on the ratio of tin to lead. You need to add about 100° to the figure to melt the solder well, another 150° for printed-circuit board connections or another 200° for conventional connections, to make up for heat carried away by the metal of the connection. This adds up to a little over 600°F as the minimum to be transferred from the tip. This still does not take into consideration the drop in tip temperature as heat is drawn from it. There must be reserve heat behind the tip, and there must be good recovery from the heating element. Too high a temperature could damage semiconductor components or lift copper from a printed-circuit board. High temperatures can also char or ruin the effectiveness of the flux before it can do its job of reducing the oxide on the surface to be soldered. Too low a temperature will take too long to transfer heat from the terminal to the solder and melt it, and the heat has a chance to affect attached parts that can't take the extra heat.

In actual practice, the lowest heat for a printed-circuit board connection is about 550°F at the tip; 1000°F is the top end for average electronic terminals. Large wire terminals may require more than 1000°F for a fast job, for example, making solder connections directly to a metal chassis. Here are the wattage ratings of the smaller pencil irons, shown with the approximate temperatures you can expect at the tip of each iron.

Watts	*Temperature*
25	650– 700°F
40	750– 800°F
50	850–1000°F

Heat Recovery

An important factor in industrial use of soldering irons is the recovery factor: at what temperature does the iron maintain a continuous supply of heat for rapid soldering? Fig. 4-11 is the recovery chart of one manufacturer. Nearly all manufacturers supply such charts where needed. The chart shows the temperature at which the iron is first able to melt solder (A on the chart). The curve reaches a maximum point and tops off at the temperature of heat equilibrium (point B). The horizontal line at the right in the curve is the temperature the iron can maintain under continuous soldering use, such as on a production line.

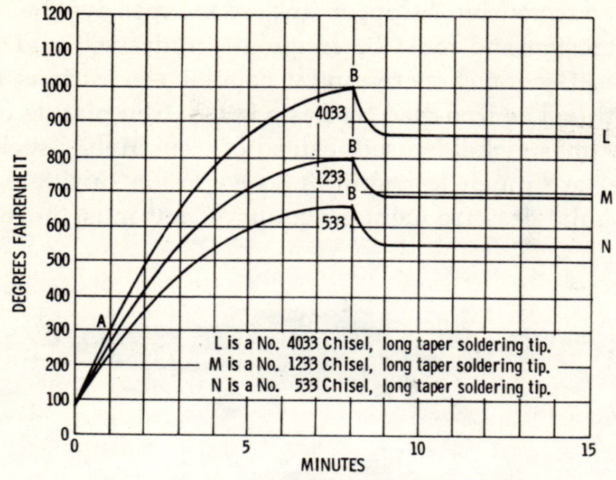

Courtesy Ungar Division of Eldon Industries, Inc.

Fig. 4-11. Recovery chart of an Ungar general-purpose soldering iron.

TIPS

With a few exceptions, tips are replaceable on all soldering irons. In very low cost "dime-store"-type soldering irons, nothing is replaceable. In some miniature irons, the tip is an integral part of the screw-in heating element, the object being interchangeability of both heating element and tip, depending on the type of soldering being done.

As mentioned before, some tips are solid-shank and are held by setscrews or nuts, in the larger irons. Most miniature-iron replaceable tips are screw-in. Fig. 4-12A shows styles of interchangeable tips in screw-in heating elements, which are themselves interchangeable. At the top is a straight shank 1/8 inch in diameter. The lower part of Fig. 4-12A shows a stepped shank tip, resulting in a diameter of only 1/16 inch at the tip. Fig. 4-12B is a close-up of two styles of the stepped tip.

One reason for tip interchangeability is so that a new tip can be installed when the old one wears out. The iron itself has almost indefinite life, but the tip is subject to oxidation and corrosion and will need replacement, some tips sooner than others. Another reason is so that the style of tip can be changed when the work to be soldered calls for it.

Most tips are made of copper because copper is the second best conductor of heat. (Silver is the best, but a solid silver tip would be much too expensive.) Copper tips are inexpensive, but are subject to corrosion and must be frequently redressed and retinned during use. Becoming increasingly popular are various forms of iron-clad tips. The iron-clad tip has a heavy iron plating over copper. Sometimes nickel is used instead of iron. Metals such as iron and nickel are much less subject to corrosion during soldering, and so require less frequent retinning (they must never be redressed).

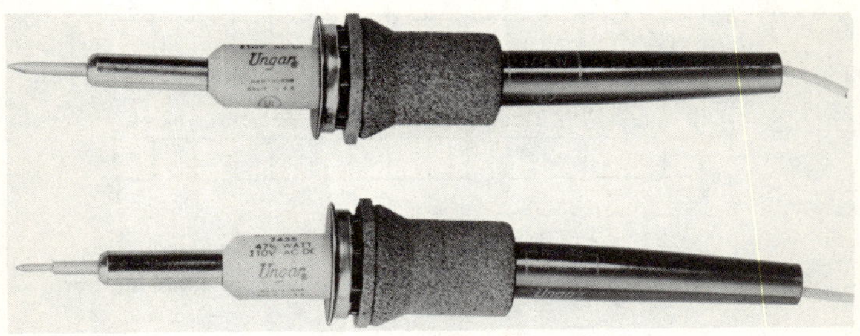

(A) Interchangeable tips.

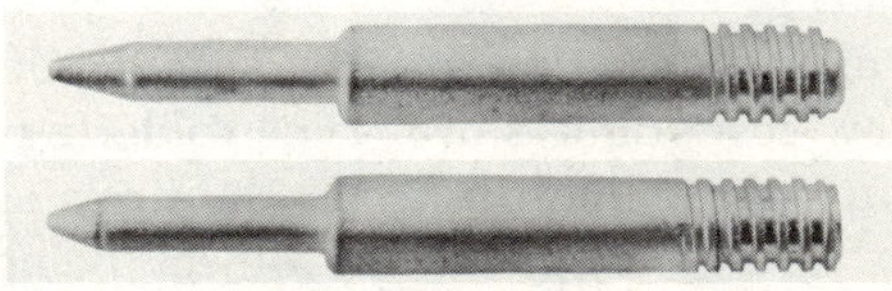

(B) Stepped tips.

Courtesy Ungar Division of Eldon Industries, Inc.

Fig. 4-12. Heating elements and tips.

Tip Sizes and Contours

The choice of tip governs the choice of iron, and is dependent on the type of soldering to be done. The object is to quickly supply the right amount of heat flow to a connection. A 100-watt standard iron with a $\frac{3}{8}$-inch-diameter tip and the chisel shape shown in Fig. 4-13 is excellent for electrical soldering on anything other than printed-circuit boards. It will handle just about any type of larger electrical connection, even soldering to a chassis. Maximum contact with the metal to be soldered governs the shape or contour of the tip. The pyramid is also a good universal shape.

For soldering electronic circuits having miniature components, the iron just mentioned is too large and clumsy. It also develops too much heat for small connections, and so requires much more careful control. One of the miniature pencil irons with about a

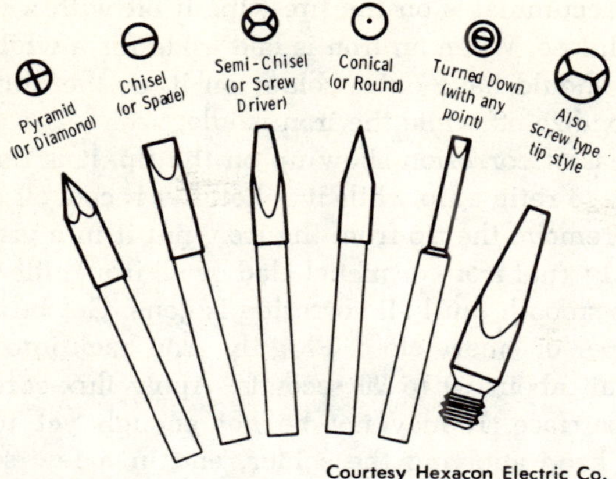

Courtesy Hexacon Electric Co.

Fig. 4-13. Tip contours most commonly used in soldering.

30- to 40-watt rating is best for smaller connections. A tip with a ⅛-inch- or $\frac{3}{16}$-inch-diameter shank is good for this size iron, and may also be used for printed-circuit boards (Fig. 4-14). For the very tiny connections on some of the very small board circuits, a $\frac{1}{16}$-inch-diameter tip in about a 20-watt iron may be easier to use. The most common tip shape in the tiny sizes is conical.

Tip Care and Retinning

Corrosion on a tip is an insulation barrier and will prevent the proper transfer of heat from the iron to the connection. A well-tinned tip is the best way to get a good and quick transfer of heat.

Tinning is the commonly used word for a coating of solder. There should always be a thin layer of bright solder on the shaped part of the tip, the part used to make the connections. When ex-

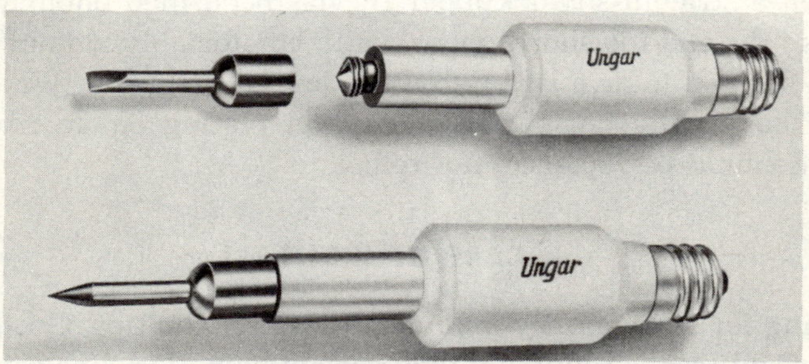

Courtesy Ungar Division of Eldon Industries, Inc.

Fig. 4-14. Interchangeable heating elements and tips with ⅛-inch-diameter shanks.

cess solder accumulates on the tip, wipe it off with a wet sponge made of cellulose. When an iron is laid aside for a while but kept hot, the tip should have extra solder on it, or the normally thin layer will oxidize off while the iron is idle.

When areas of corrosion show up on the tip, it is time to retin it. Never try to retin a tip while it is hot. Let it cool off completely first. Then, remove the tip from the iron, put it in a vise, and if it is copper only (not iron- or nickel-clad or all iron), file the soldering surfaces smooth until all corrosion is gone. Get burrs off with fine sandpaper or emery cloth. Plug the iron back into the power line, and wait about 60 to 90 seconds. Apply flux-core solder to the bright surface. It may not be hot enough yet to melt the solder, but keep applying the solder, and in a few seconds the solder will begin to melt. Retinning should be done at the coolest tip temperature possible; otherwise, corrosion begins to set in fast at higher temperatures, and tinning will be more difficult. As the iron warms up, apply solder to all surfaces. When covered, wipe off the excess on a wet sponge.

The iron-clad and similar tips should never be filed. Since they resist corrosion much better than copper, it will take only a few swipes of an emery cloth to restore the surface to a brightness that will take a new retinning.

Although the tips of instant heat guns are made of copper, they must never be filed. Filing copper off these tips will increase their ohmic resistance, thus reducing developed heat. A light sanding will restore their surface. If these tips become pitted, they should be replaced.

When you are through using an iron, lightly wipe off excess solder as the iron cools. Always leave a liberal but clean coat of solder on the iron.

When a straight-shank copper tip has been filed enough times to make the end too short to work with comfortably, do not pull it out of the iron to give it length. It is time to replace it. When corrosion has finally cut through the iron plating on an iron-clad iron, it should be replaced, not refiled.

CARE OF IRONS

Caring for irons mostly involves caring for the tips. Good-quality irons will last a lifetime if they are not abused. A line cord may wear out on occasion, but replacement is easy in most irons.

Since the tip is part of the radiating system that establishes a safe temperature equilibrium, an iron should never be left on without the tip installed. Otherwise, the internal heat may rise to a temperature too high for long element life.

When excess solder accumulates on a tip, wipe it off on a wet cellulose sponge; never hit your iron against the workbench. When the iron has cooled off, remove the tip. This will prevent the tip from freezing to the iron and thus prevent difficult removal later on. On hollow-core standard irons, a light tap (with the tip out) will knock out internal corrosion particles.

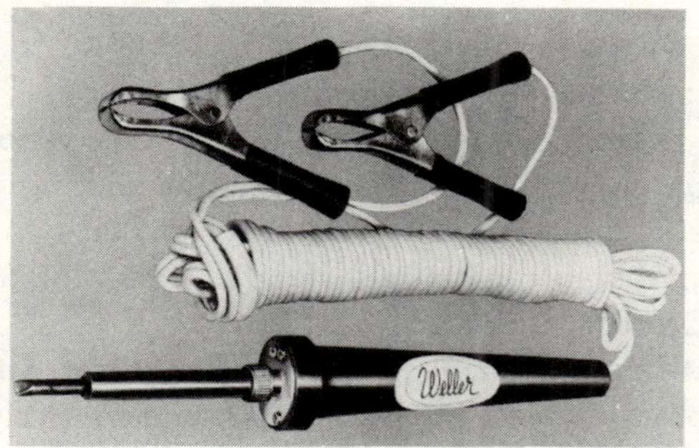

Courtesy Cooper Industries

Fig. 4-15. Weller iron equipped with clips for operation from car storage battery.

CAR-BATTERY-OPERATED IRON

The Weller iron in Fig. 4-15 has a 12-volt heating element, and is equipped with clips so that it can be connected to a car-battery power source. It is handy for making solder connections from the car when ac voltage is not available. The heating element is equipped with Weller's magnetic temperature-sensing control described earlier.

5

Instant-Heat Guns

The "instant-heat" soldering gun is a special type of soldering iron that is on only when you are actually making a solder connection. When you need to solder only one or two connections, you can use an instant-heat gun instead of a regular soldering iron, which takes 5 to 10 minutes to heat up.

The instant-heat soldering gun has a very low duty cycle (on time to off time), which means it uses considerably less power than an iron that is on continuously. This feature is a useful one in television service shops, where irons need to be ready for use but need not be on all the time. Electronic-kit builders and home experimenters will find this type of iron very handy.

Another feature of the instant-heat gun is its efficiency. The gun generates the heat at or near the point of soldering. Because of its long off time, it does not need a heat-radiating design to obtain and maintain heat equilibrium, as do regular irons. So, less heat is radiated away as waste heat.

The word "gun" comes, of course, from its shape and the location of the on-off trigger switch. In spite of the fact that its built-in transformer makes this gun heavier than conventional irons, the gun shape does give it good balance in the hand. The weight, however, could be fatiguing if the gun were held constantly, as in line production work.

HOW IT WORKS

Heat is a function of power, and power in watts is volts times current according to the formula $P = EI$. You can get the same power from high voltage and low current, or high current and low voltage. For example, 120 volts times 1 ampere equals 120 watts (IE = P; 120 × 1 = 120), or 1 volt times 120 amperes equals 120 watts (1 × 120 = 120). In a regular soldering iron, the 120-volt ac line voltage is applied to the ends of a wire coil, whose resistance is selected to develop the power desired. The instant-heat gun has a transformer in the housing that steps the 120-volt line down to less than 1 volt. A single-turn loop of copper is the soldering tip and, when connected to the secondary of the transformer, carries several hundred amperes of current, depending on the resistance of the tip and the voltage output of the transformer (Fig. 5-1).

Fig. 5-1. In instant-heat gun, high current heats thin bend of the tip wire.

In instant-heat soldering guns, the tip is the coil, called a single-turn coil. A piece of copper is bent like an oversize modified hairpin and connected to the transformer secondary. The current in this piece of copper heats the copper to soldering temperatures. Being copper, it is used to solder with directly. Because there is very little time lapse (about 3 seconds) between the time the current is applied and the time the tip is hot enough to solder with, it is given the name "instant heat."

SELF-REGULATION

Copper has a rather high coefficient of expansion, higher than that of the nichrome wire used in conventional soldering irons. When the gun is first turned on, the current in the soldering tip is quite high, but quickly drops down as it gets hot, due to an increase in its resistance. When the tip is touched to a connection to be soldered, some of the heat is, of course, drawn out of the tip. This cools the tip a little and decreases its resistance, which results in an increase in current, and, therefore, more heat. This is self-regulation.

The tips are bent into long hairpin pieces to reach into chassis corners or tight places. Some are attached to brass posts extending out as parallel connectors from the transformer housing, or brazed to a length of heavier brass at each end. The heat is generated in the short copper section and is fairly well confined at or near the soldering end. On some, the copper is thinner at the tip, so there will be higher resistance at the tip, which is necessary for concentration of heat there.

Why aren't copper coils used in regular soldering irons? The reason is that copper oxidizes, and at high temperatures the corrosion of copper is accelerated. When a coil of wire is confined to the inside of a protective casing, it can become quite hot, and it will not have long life. In the case of the single-turn tip on an instant-heat gun, the wire is exposed and is on only a few seconds at a time, so corrosion is kept at a minimum. Even so, the instant-gun tips do need occasional replacement, but fortunately they are inexpensive.

A coil of copper wire is used in a modified version of the instant-heat gun. It will be described later in this chapter.

MANUFACTURERS OF INSTANT-HEAT GUNS

There are not many manufacturers of instant-heat guns. The two principal companies are Cooper Industries, manufacturers of the Weller iron (Fig. 5-2), and Wen Products (Fig. 5-3).

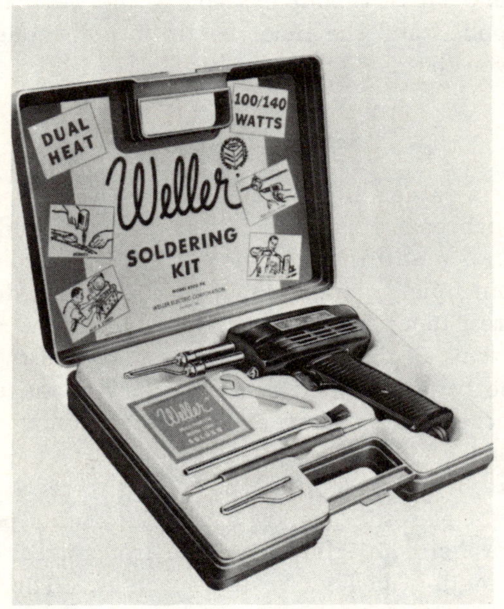

Courtesy Cooper Industries

Fig. 5-2. Weller instant-heat soldering gun.

There are two Weller models. Each features dual-heat ratings of 100/140 watts and 240/325 watts. A tapped transformer supplies either a high or a low voltage to the tip. A light squeeze on the trigger switch produces low heat; a heavier squeeze produces high heat.

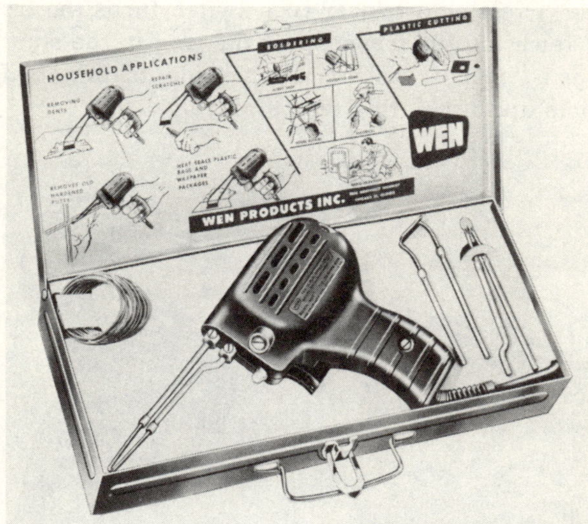

Courtesy Wen Products, Inc.

Fig. 5-3. Wen instant-heat soldering gun kit with three tips, one for soldering, one for cutting plastic, and one with a flat-iron tip.

There are five models of the Wen iron, each with a progressively higher wattage rating: 75 watts, 100 watts, 130 watts, 200 watts, and 450 watts.

All of the Weller and Wen instant-heat guns have small bulb lights to help illuminate the work being soldered.

ACCESSORY TIPS

Interesting accessory tips are available for instant-heat soldering guns. A tip may be purchased having a flat end for sealing thermoplastic material or for taking dents out of wood. For cutting through vinyl tile and for wood and leather burning (Fig. 5-3), a tip having a sharp cutting edge may be purchased. The tips on the guns are easily replaced by loosening setscrews or knurled nuts.

THE COMPROMISE GUN

Wen Products, Inc., has developed a compromise soldering gun that uses a heating coil and soldering tip like those on other soldering irons (Fig. 5-4). The small, high-current coil is located very near the tip. It operates at low voltage from a step-down transformer. Pressing a trigger-type switch turns the iron on for soldering; releasing the trigger puts the iron in the standby position. This is a "nearly" instant heat scheme, taking about 10 seconds to come up to soldering temperature.

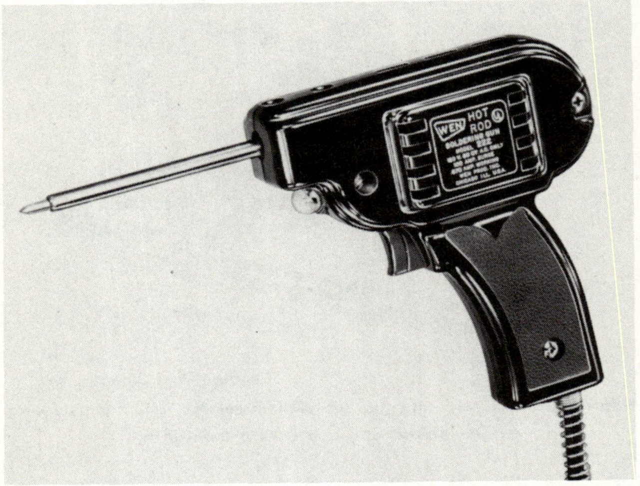

Courtesy Wen Products, Inc.

Fig. 5-4. Wen Model 1222 soldering iron.

The coil is made of copper, so a degree of automatic temperature control is achieved in the same way that it is for the single-turn tips. The result is an increase in resistance of the coil when it is hot. The copper has a noncorrosive metal coating and a fired-on ceramic covering to prevent corrosion of the copper.

Another soldering gun, but one based on a different principle, is the Wall "Trig-R-Heat" (Fig. 5-5). It does not use a transformer. The full 120 volts ac is applied to the heating element in much the same manner as it is to a standard soldering iron. However, the element develops very high wattage. Squeezing the trigger switch applies power, and it takes only a few seconds for the tip to come up to soldering temperature. When it does, the trigger

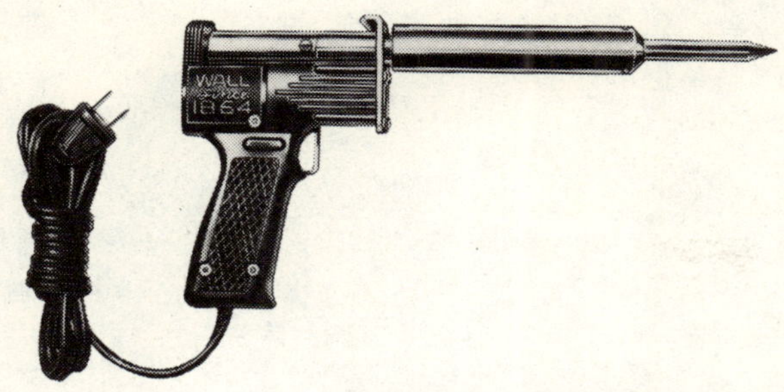

Courtesy Wall-Lenk Manufacturing Co.

Fig. 5-5. The Wall "Trig-R-Heat" soldering gun does not use a transformer.

should be released. With experience, you can maintain the desired soldering temperature with the on-off manual switch.

There are two Wall soldering guns. One has a 400-watt element and a ¼-inch tip; the other has a 550-watt element and a ½-inch tip.

6

Cordless Soldering Irons

The nickel-cadmium battery made the cordless soldering iron possible, just as it made numerous other electric-powered devices possible. The electric toothbrush, cordless electric shaver, and powered grass shears are just a few such devices. Most of these devices use electric motors to do their work. The nickel-cadmium battery (Nicad) is a small, lightweight rechargeable battery, capable of delivering fairly heavy amounts of current when needed. It is especially suited to hand-held equipment that gets only occasional use. All types of equipment using the Nicad battery are supplied with chargers that can recharge a dead battery overnight. The chargers are physically small, and are often the holders for the devices.

Occasional soldering jobs are excellent uses for cordless irons. The small battery in the iron can supply enough heat to make from 50 to 100 soldering connections per charge, depending on the size of the wire and the connections. The irons are small, easily held like a pencil, and have small tip sizes, which make them excellent for printed-circuit board work.

ADVANTAGES OF THE CORDLESS IRON

There are many advantages to using a cordless soldering iron. Its independence from connection to an ac outlet means greater portability. With no line cord dangling from the device, it is easier to work in tight places. When working with certain types of semiconductors (MOSFETs and other similar types), it is important that no electrical leakage reach the high-impedance input gate, or damage could result. Although there are special soldering irons that incorporate devices to isolate them from the ac line, the cordless iron is automatically isolated.

All the cordless soldering irons described in this chapter are very much alike. They are similar in appearance, have small lights for work illumination, and have locking devices to prevent accidental switch activation.

A DIFFERENT SOLDERING TIP

In the cordless soldering iron, the heating element is encased in the tip. Thus, the heat for soldering is developed inside the tip. Tip current is supplied from a battery (approximately 3 volts). This current passes directly to the heating element in the tip. Soldering temperature is reached in seconds, which eliminates the need to leave the iron on all the time as is necessary with conventional irons.

Some manufacturers make a variety of tips, each best suited to a particular type of soldering. Some of these tips are secured with setscrews and others plug in.

THE CHARGER

All cordless soldering irons come with a small charger. In some models, the charger is the holder for the iron; in others, it is a separate, plug-in device. The charger consists of a small transformer and rectifier, fully enclosed. On the average, it supplies 3 volts at 120 mA to a completely dead battery. The current decreases as the battery comes up to charge.

BATTERY CARE

Nicad batteries require a certain amount of care. They should always be left on charge when not in use and should never be allowed to remain in a discharged condition for long periods of time. After you have finished your soldering work, the iron should be put back on charge and left there until the next time it is to be used. Even a fully charged battery, if left for a time without being on charge, will discharge a certain amount through leakage. Before you use a battery that has been in storage for three months or more, or before you use a newly purchased rechargeable iron (which may have been on the dealer's shelves for some time), you should put the iron through three or four charge-discharge cycles. The Nicad battery is built to be charged and discharged. Manufacturers say that their Nicad batteries will last through 500 to 1000 discharge-charge cycles.

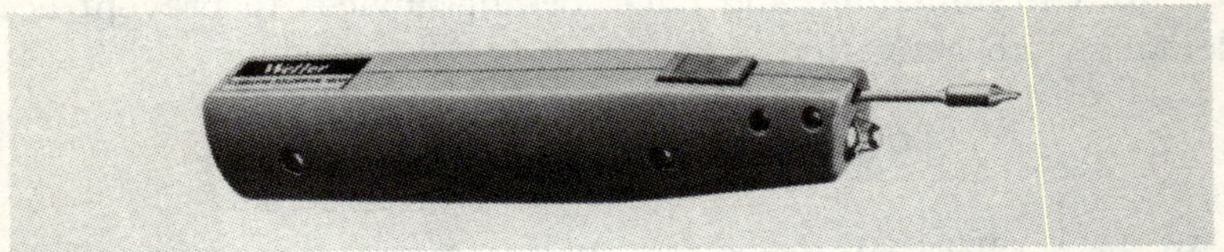

Courtesy Cooper Industries

Fig. 6-1. The Weller cordless iron is typical of most. When not in use, it is hung up by a ring on the back.

THE WELLER CORDLESS IRON

The Weller cordless iron is fairly representative of most cordless soldering irons (Fig. 6-1). Its shape, size ($6\frac{7}{8}$ inches long without tip), and weight ($5\frac{3}{4}$ ounces) make it easy to handle during soldering. It has a prefocused bulb that provides work illumination and has a locking device on the handle of the iron that prevents accidental discharge of the battery. Four interchangeable tips (Fig. 6-2) are available for this iron. Three of the tips are about $2\frac{1}{2}$ inches long with varying shapes; the fourth is $3\frac{1}{2}$ inches long and is used for reaching into tight spaces such as tv tuners. The tips reach soldering temperature in 6 seconds, developing 10 watts of heat power. The tips are nickel-plated copper and come pretinned, ready for use.

The charger (Fig. 6-3) is small in size, and plugs directly into an ac outlet. One end of the cord is attached to the charger; the other end has a phono-type plug that plugs into the iron. Instead of using the charger for a stand, this iron has a loop for hanging it on a wall or the side of a workbench. The charger has an output of 2.9 volts at 120 mA and will recharge a completely dead battery in 14 hours.

UNGAR CORDLESS IRON AND STAND

The Ungar iron (Fig. 6-4) is similar in appearance and operation to the Weller iron, but it has a few interesting features. In

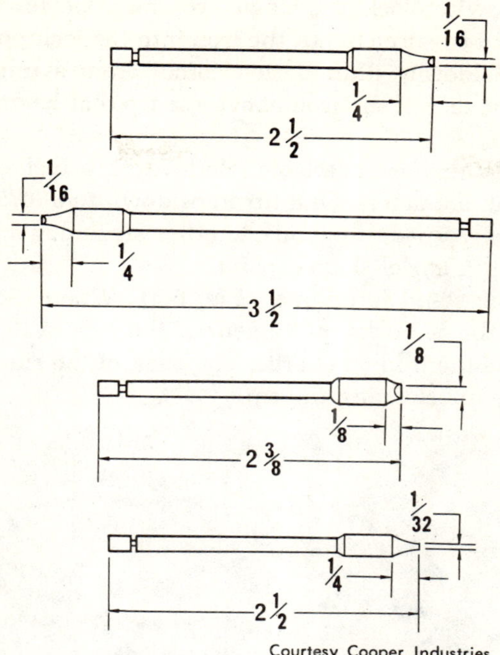

Courtesy Cooper Industries

Fig. 6-2. Interchangeable tips for use with the Weller cordless iron.

Courtesy Cooper Industries

Fig. 6-3. The Weller charger plugs directly into an ac outlet. A long cord with plug fits a jack on the back of the iron.

addition to a switch lock, the Ungar iron has a power-on indicator light to remind the user to put the iron into the lock position when not in use. Soldering illumination comes from a tunnel-like depression at the end of the iron above the tip that beams light onto the work area.

Two plug-in tips are available. Both are 1⅞ inches long, with 3/16-inch shank diameters. One tip steps down to a .075-inch diameter with a microspade tip, and the other steps down to a ⅛-inch diameter with a chisel-shaped point.

The charger stand is unique (Fig. 6-5). When the charger is plugged into the ac outlet, placement of the iron in its stand automatically puts the iron on charge. The back of the stand has a cellulose sponge for cleaning the tip.

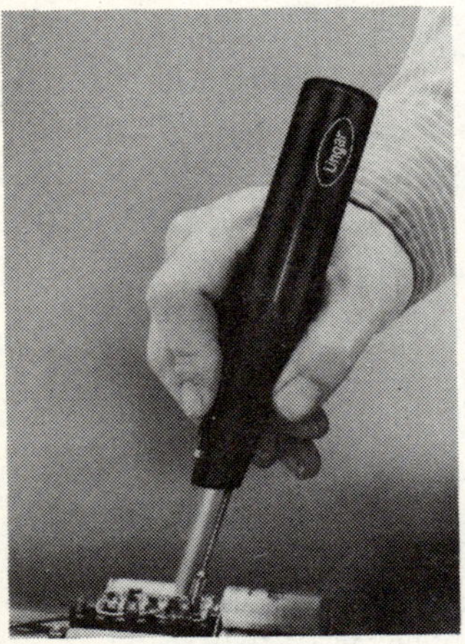

Courtesy Ungar Division of Eldon Industries, Inc.

Fig. 6-4. This iron includes a power-on light to remind the user to place the iron back on charge when it is not in use.

Courtesy Ungar Division of Eldon Industries, Inc.

Fig. 6-5. The Ungar stand includes a cellulose sponge for tip cleaning.

THE CORDLESS WAHL WITH DRILL ATTACHMENT

Probably the iron with the greatest versatility and the one capable of developing the highest heat is the cordless Wahl iron made by Wahl Clipper Corp. The iron (Fig. 6-6) develops a temperature of 700°F and dissipates about 50 watts of heat power at the tip. There are 10 plug-in tips with different shapes (Fig. 6-7), plus one extra-long tip for television tuner or similar hard-to-reach places. The iron is 8 inches long and weighs 6 ounces.

Fig. 6-6. The Wahl cordless iron in its charging stand. The stand has space for stowing other tips.

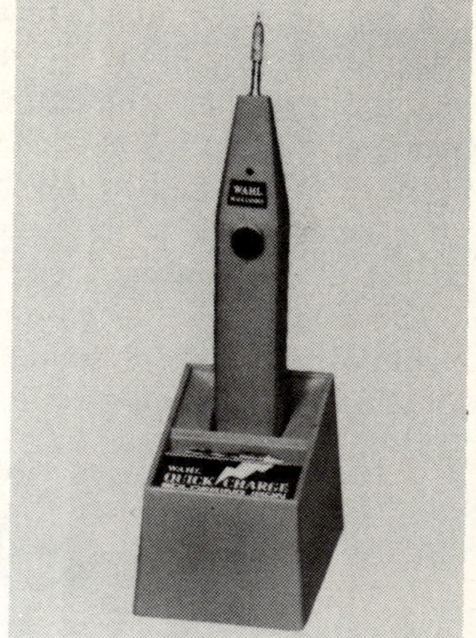

Courtesy Wahl Clipper Corp.

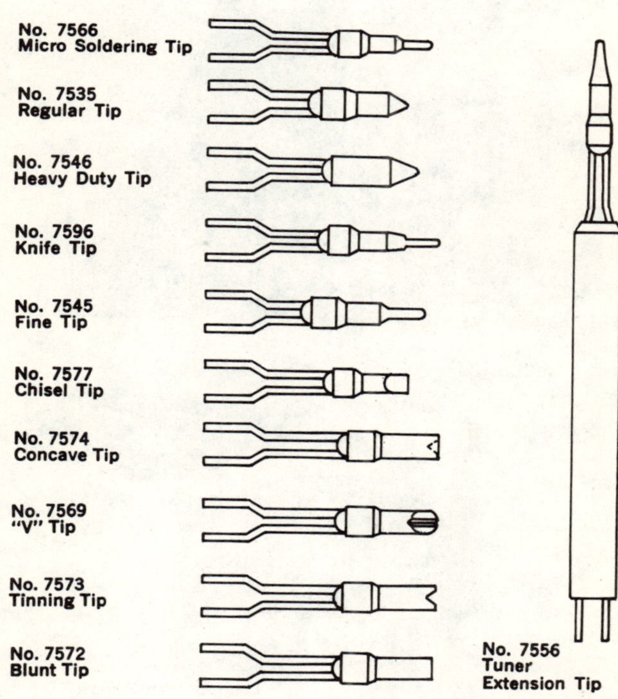

Courtesy Wahl Clipper Corp.

Fig. 6-7. The Wahl iron has the largest choice of tip shapes of any of the cordless irons.

A unique and valuable feature of this iron is the accessory drill attachment (Fig. 6-8). By removing the soldering tip and plugging in the drill attachment, you have a cordless drill with a small-diameter bit that whirs at 14,000 rpm under no load. It is ideal for opening up printed-circuit board holes that may have filled in after removal of a component part. This method reduces the chance of overheating the printed-circuit board copper.

The charger is a bench-top stand that holds the iron in an upright position while charging. A 12-volt charger adapter that plugs into an automobile cigarette lighter socket is also available.

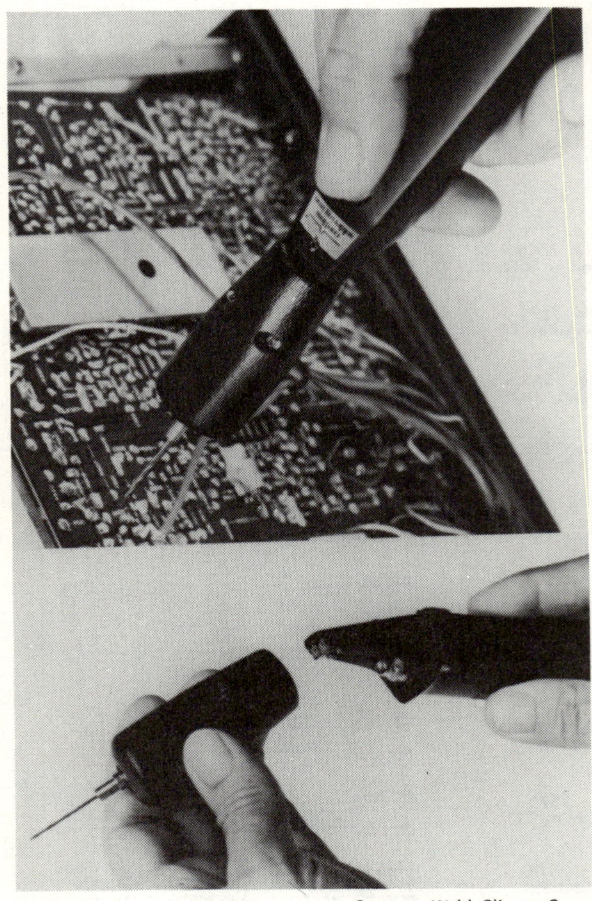

Courtesy Wahl Clipper Corp.

Fig. 6-8. A drill attachment plugs into the Wahl iron. Its tiny bit is ideal for opening holes on a printed-circuit board .

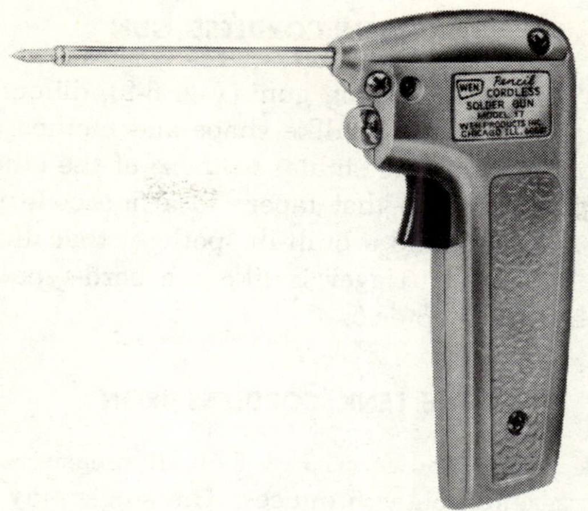

Courtesy Wen Products, Inc.

Fig. 6-9. The Wen cordless soldering gun.

Courtesy Wall-Lenk Manufacturing Co.

Fig. 6-10. The Lenk cordless soldering iron is a single-tip tool. The charger is a plug-in type with extension cord to the iron.

THE WEN CORDLESS GUN

The Wen cordless soldering gun (Fig. 6-9) differs from other cordless irons only in its gunlike shape and method of handling. Otherwise, its features are similar to those of the other irons discussed. It has a long tip that tapers to a fine cone for reaching hard-to-get-at places, and a built-in spotlight that illuminates the work area. Its switch trigger is like the cord-type instant-heat guns described in Chapter 5.

THE LENK CORDLESS IRON

The Lenk iron is shown in Fig. 6-10. It measures 8 inches in overall length and weighs 6 ounces. The single tiny tip heats to 700°F, reaching this temperature in 5 seconds. The manufacturer suggests that by removing the tip you can use the built-in work light as an emergency flashlight.

The charger plugs into a 120-volt ac socket. One end of the cord is attached to the charger, and the other end of the cord plugs into the iron. With this type of charger, soldering can be done while the iron is charging.

7

How to Solder

This chapter is about soldering with a soldering iron. Other systems of soldering are dealt with in chapters that follow. Whether you are a beginner or an experienced solderer, there is much of value here for you. The information is fundamental but complete.

SOLDERING SUPERIORITY

Solder guarantees positive electrical continuity, and it "locks" mechanical connections securely. However, solder is soft and cannot be depended on to hold two pieces of metal together by itself in a long-lasting and secure connection. By itself, solder has rather poor holding power, and can be pulled apart with little effort. This means, of course, that there should first be a good mechanical connection. Wires should be wrapped around a terminal post, or put through a terminal hole and wrapped around the terminal.

A wraparound connection by itself is also a poor connection. Wire is pliable and can become loose. Furthermore, the thin oxide coating that forms on copper and other normal soldering metals can introduce resistance in the connection. If it is a connection that passes heavy current, there will be enough resistance to affect the performance of the circuit. The answer is a combination of a good mechanical connection and proper soldering.

In a properly soldered connection, the flux dissolves away the oxide, and the solder forms an intermetallic bond with the metals being soldered. The hardness of the solder, combined with the hardness of the wires, sets up a strong mechanical tie that is much better than each separately. There are exceptions to the combination. One, for example, occurs when a low-resistance ground is being provided for a variable capacitor. Often a heavy braided wire is soldered between the variable capacitor frame and the chassis to which it is mounted. Of course, the only mechanical strength called for is to hold the braid in place. Another exception, in modified form, is the soldering of components to a printed-circuit board. While component leads are pulled back to hold the component in place, the real holding strength is in the solder. However, such parts are small and light, and not much mechanical strength is needed.

THE RIGHT IRON AND TIP

The soldering iron has one purpose: to heat the connection so that applied solder and flux will melt onto the connection. But the right amount of heat is one of the prerequisites for getting good solder connections. Since the ideal temperature is hardly ever achieved, the object is to get as near to it as possible.

The ideal temperature for a soldering iron is found as follows: It takes about 370°F to begin to melt the usual solders. It takes another 100° to be sure it is in a liquid state. It takes another 150° to 200° at the tip of the iron to allow for fast drain-off of heat as the iron is applied to the connection. This adds up to about 620°F. This tip temperature is adequate for the smallest connections.

From a practical standpoint, 620°F should be a minimum temperature for point-to-point soldering. Soldering components to printed-circuit boards can be done effectively with less heat, although 600° to 700°F at the tip is better for making quick printed-circuit board connections. This avoids the risk of heat being carried along the copper foil when insufficient heat is used and longer time is needed to make the solder connection. Soldering a connection having a large mass of metal tends to draw heat from the iron rapidly. This means the iron must have a high wattage rating and plenty of copper in the core of the heating element. Furthermore, making solder connections in rapid succession calls for an extra reserve of heat from the iron. Production solderers need a hotter iron so that they can make connections more quickly, and thus minimize the amount of heat that has a chance to travel along the leads of the components and possibly damage the components, or in the case of printed-circuit boards, the copper foil. So, starting with the optimum tip heat of 650° to 680°F, you add more heat depending on the factors mentioned. Large connections, with their fast drainoff of heat, could require a tip temperature as high as 1000°F. Too much heat, though, will char the flux and make it ineffective, resulting in a poor connection.

Because it is important to transfer heat from the iron to the connection as fast as possible, the tip contour must be designed to present the maximum surface to the connection. A pyramid or modified chisel shape is best for average use, particularly in the home or shop. There are many special shapes, but they are used mainly in production soldering where the shape must fit the specific connection.

Lastly, the condition of the tip is important. It must be well tinned with a bright coat of solder, and have no corrosion spots or pits. A good coat of solder is the best way of transferring heat quickly from the iron to the connection. (See Chapter 4 for methods of tinning a tip.)

SOLDER AND FLUX

Except for special applications, the solder used should have a core flux and should have one of two tin/lead ratios. The 60/40 and 63/37 ratios are the two best for regular soldering. The 63/37 ratio is the eutectic ratio. That is, this solder goes from a liquid to a solid without going through the "pasty" stage. The 60/40 ratio is just about as good. It is so near the eutectic ratio that the pasty stage is only a few degrees between solid and liquid. The pasty stage is no big problem, anyway. It just means that the connection must not be disturbed while cooling until it has solidified, or a poor connection will result. Both of these solders have a high ratio of tin to lead. This means that they have a greater ability to wet the metals being soldered and will make a stronger bond than solders with lower tin ratios.

Flux-core solders come in different wire sizes or diameters. Size 20 or 0.036 inch is good for printed-circuit boards as well as other connections. If you work exclusively with equipment other than printed-circuit boards, a larger size, about 16 gauge or 0.064 inch, feeds slower and may be easier to handle. Stay with flux-core solders; a separate flux is rarely needed.

CLEANLINESS

As you are soldering, the flux in the solder will clean off the thin oxide coating that forms on copper wire or tinned leads. However, it will not remove grease, paint, heavy corrosion from years of storage, or the skin oils from sweaty fingers. For the solder to do a good job of wetting, the metals must be clean and bright.

Brand-new components are clean and in solderable condition. There is nothing you must do to them except keep your fingers off the leads as much as possible. Handle the parts by their bodies: hold printed-circuit boards by the edges, as you would a phonograph record.

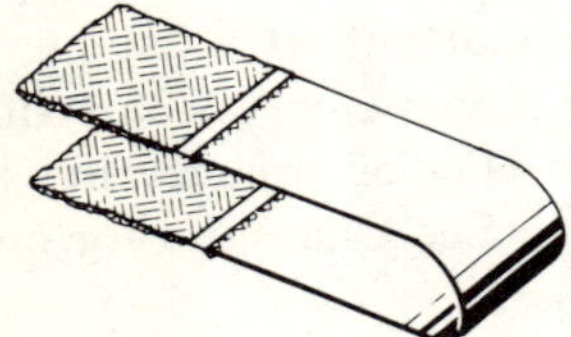

Fig. 7-1. NASA-approved scraper for cleaning component leads before soldering.

Parts that have been in a dealer's stock for a long time, or that have been in your "junk box," could have a heavy coat of oxide on the leads. Clean them with a few swipes of fine sandpaper or emery cloth. The NASA agency for aerospace experiments recommends a homemade gadget like the one illustrated in Fig. 7-1. It is made of ½-inch tinned copper shielding braid. Two pieces are used. One is folded over each end of a U-shaped piece of spring metal. The edges may be soldered on, or a clamp arrangement may be used. Squeeze the ends over the component leads and draw the leads out. Do this two or three times as you revolve each lead. Oil films, grease, and paint will need to be thoroughly cleaned off with a suitable solvent.

WIRE STRIPPING

The ends of a hookup wire should be stripped of their insulation. Production lines have automatic machines for this. These automatic machines are expensive and impractical for service technicians or hobbyists, though. Service technicians and hobbyists will find it worthwhile to invest in a hand stripper. One such device holds the wire behind the part to be stripped, cuts the insulation, then uses a pair of claws to grip the insulation and pull it away from the wire.

A simpler, very inexpensive hand stripper is also available. This type depends on your holding the wire while it strips the insulation. It is easy to use and does a good job of stripping the insulation from various sizes of wire without damaging the wire.

A third type of hand stripper is the popular crimper/stripper shown in Fig. 9-6 of Chapter 9. Its versatility makes it an attractive choice.

If you do not have a wire stripper, you can still strip the insulation off wires. A sharp knife will do it, but it is a tedious job. One of the best methods is to use diagonal wire cutters. Cut a circle around the insulation with the cutter; then with the cutter in the circle pull the insulation off (Fig. 7-2). With practice you can learn to cut nearly through the insulation without nicking the wires.

The insulation should be cut back to about 1/8 inch beyond the calculated point of solder. If the insulation ends near the connection, some of the plastic material could melt into the connection and prevent good wetting by the solder. If it is too far away, you may lose the protection of the insulation against shorts with other wires nearby.

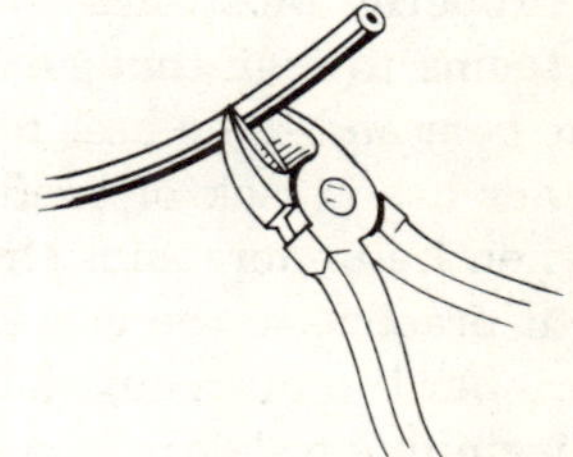

Fig. 7-2. Wire stripping with diagonal wire cutters.

PRETINNING LEADS

You may have observed the ends of the stranded leads on a power transformer or iron-core choke. They have been twisted tightly together and tinned. At the factory, the ends are first stripped of insulation, then dipped in a liquid flux, and finally put into a pot of molten solder for a couple seconds to fuse the ends together like solid wire. This makes them easier to insert into terminal holes and bend around a terminal, and also makes them easier to solder.

You may find it more convenient to work with pretinned leads, especially if they are stranded leads. It is easier to tin leads by handling the stripped wire and solder and leaving the soldering iron lying on the workbench (if it's the miniature type with a flange that permits this without a regular stand), or chucked in a vise. Place the bared, tightly twisted, stranded wires against the iron for a second, then apply solder to the top side. When the wire has become hot enough, the solder will melt and run into and between the strands of wire (Fig. 7-3).

Fig. 7-3. Pretinning stranded wires.

SOLDERING FOR BEGINNERS

Manufacturers say that in 90 percent of the kits returned to them as being defective, the defect is due to poor soldering. Poorly soldered wiring connections occur when the solder does not run into all the crevices of the connecting wires and wet the metals properly. Most of this is due to insufficient heat. The beginner seems to fear that he is going to damage the connection with the heat and so he uses too little. In fact, the biggest problem a beginner has is lack of confidence. However, this is soon overcome when he understands the soldering process and gains practice—but practice in the correct methods. Bad habits learned initially are hard to overcome later on, so it is wise to take the time in the beginning to learn the right way.

Let's go through the steps of making a solder connection. Assume that you are soldering the end of a resistor lead to a terminal of a tube socket on a chassis.

Before you start, you must have the right iron (40 to 50 watts) and the correct solder for the job. The solder should have an activated flux and be either the 60/40 or 63/37 type (like Kester 44 or Ersin Multicore). The tip of the iron should be well tinned and should look bright.

Bring the iron up to temperature. Using long-nose pliers, feed the end of the resistor wire into one of the holes in the socket terminal. Twist the wire back around the rear of the terminal and

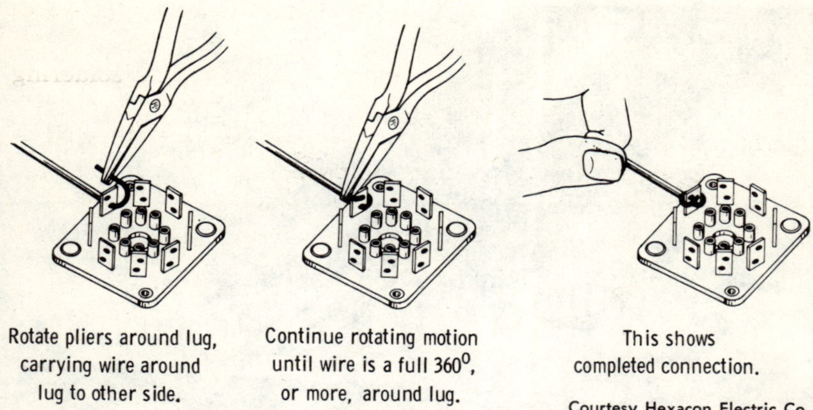

Rotate pliers around lug, carrying wire around lug to other side.

Continue rotating motion until wire is a full 360^0, or more, around lug.

This shows completed connection.

Courtesy Hexacon Electric Co.

Fig. 7-4. Best wraparound connection to tube socket.

squeeze it tightly against the terminal (Fig. 7-4). Generally, ending the lead tight against the back side of the terminal is adequate. You should always start with a solid mechanical connection.

Next, pick up the soldering iron with your right hand (left if you are left-handed). Remember that holding the iron correctly is an important part of the soldering process. Hold a miniature pencil-type iron as you would a pen or pencil. Grip a standard iron so that your four fingers wrap around the handle of the iron

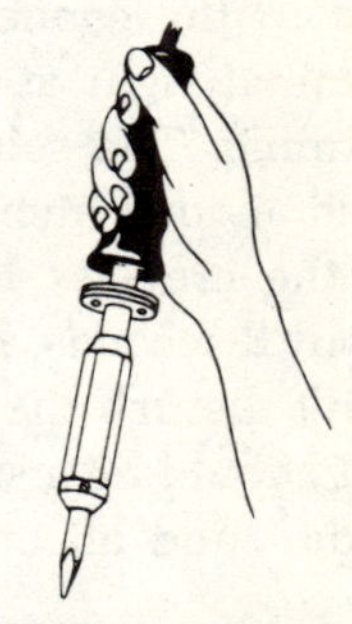

Fig. 7-5. How to hold large soldering iron for effective soldering.

Courtesy Hexacon Electric Co.

with your thumb at the cord end of the handle (Fig. 7-5). Hold the soldering gun as you would a pistol, using your forefinger to operate the trigger switch.

Take up the spool of solder in your other hand with the thumb and forefinger holding the solder about 3 inches behind the end (Fig. 7-6). Apply the tip of the iron (the flat side) to the wire and terminal. If your iron has a pyramid tip, you will be able to

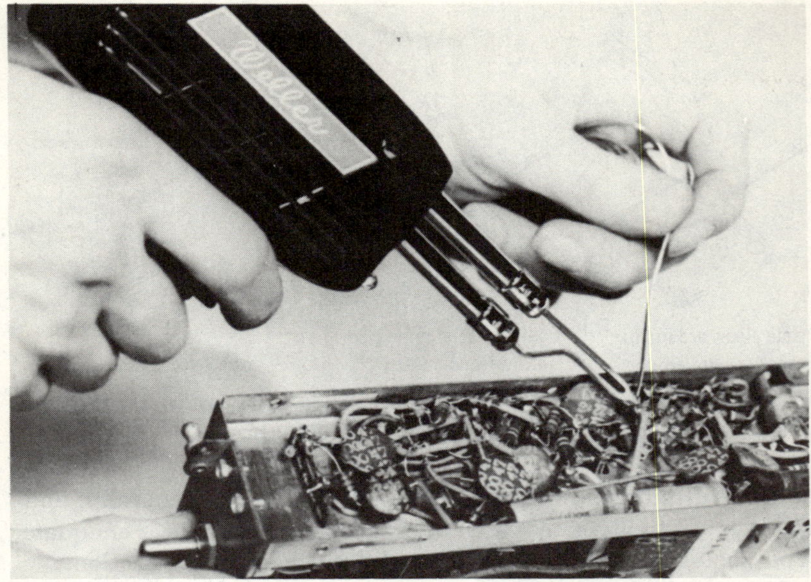

Courtesy Cooper Industries

Fig. 7-6. Applying solder to connection.

fit it against both the terminal and the wire. Don't be afraid to apply a little pressure. The object is to transfer heat from the iron to the connection. Now touch the end of the solder to the terminal right at the point where the iron is in contact with the terminal. The solder should begin to melt in less than 1 second. Feed about 1 inch of the solder into the connection. Let it flow into all the crevices between the wire and the terminal. This takes about 2 seconds. Remove the solder first, then the iron (Fig. 7-7). Don't disturb the soldered connection until it has cooled. The result should be a strong connection with a bright appearance to the solder, and all crevices filled, but with the contour of the wire

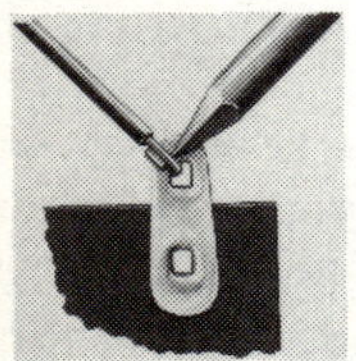

(A) Hold soldering iron against connection.

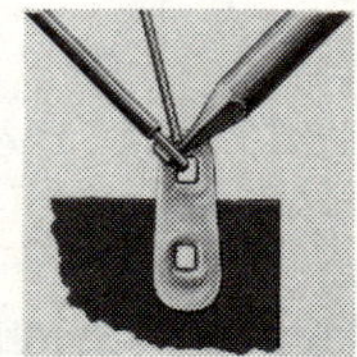

(B) Apply rosin-core solder at point of contact.

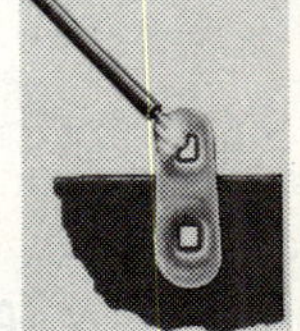

(C) Let connection cool.

Fig. 7-7. Making solder connection.

showing under the solder. The edges of the solder should feather out along the terminal and wire. Fig. 7-7C shows an example of a good solder connection.

Nothing is gained by using a lot of solder. It does not add to the strength of the connection. Use only enough solder to fill all the crevices and to cover the wire so that its contour shows through the solder covering.

As you continue soldering, it is important to maintain a bright iron. Wipe the tip of the iron once in a while with a wet cellulose sponge or a damp cloth to keep it bright.

EXPERIENCED SOLDERERS

Electronics technicians with considerable soldering experience must have read the previous paragraphs with a smile on their faces. For them soldering is an automatic process. They apply the iron, apply the solder, and take them both away, without ever actually thinking about each motion.

An understanding of what goes on during the soldering process is important, even for experienced solderers, because it will lead to better solder connections. The object of good soldering is to have capillary action promote good wetting of the base metal by the solder. Refer to Fig. 3-1 in Chapter 3 for a view of what goes on as you draw a hot iron across a piece of copper. The copper, heated by the iron, first liquefies the flux, then vaporizes it. This action dissolves the oxide coating which, invisible though it may be, is always there. All this evaporates with the heat. At the same instant the solder melts and, by capillary attraction, flows along the copper and actually intermixes with the surface layer of copper —it is in solution with the copper surface layer. There is no longer any space between the two. This is what gives the connection an electrical continuity that is even better than it was before soldering. At the interface, the copper is now in solution with the solder and forms an alloy of tin, lead, and copper.

It takes just the right amount of heat to get good flux action and good solder wetting. Not enough heating of the base metal (the copper in this example) will result in a cold solder connection with no alloying. Too much heat will burn the flux and ruin its ability to remove the oxide. Good wetting is identified by a low angle between the edge of the solder and the base metal at the edge of the solder. A rolled-over edge means poor wetting and probably poor electrical continuity between the solder and the copper underneath (Fig. 7-8).

PRINTED-CIRCUIT BOARDS

After a copper sheet has been laminated to a sheet of phenolic, the circuit may be printed, silk screened, or applied photographically, but the name "printed-circuit board" applies to all. The "printing" covers the copper foil where the electrical connections are to be. An acid dip etches the rest of the copper away, leaving only the circuit.

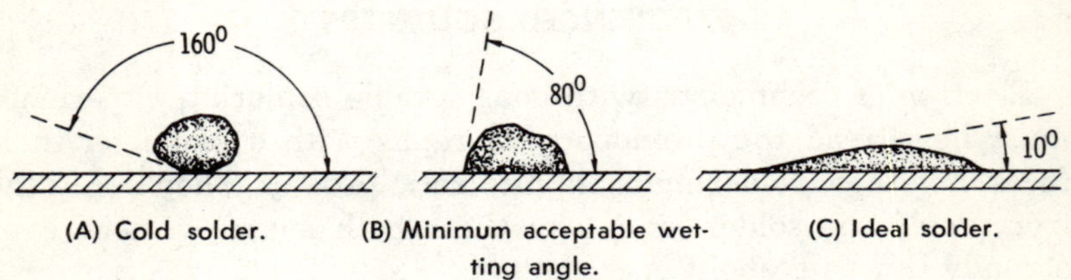

(A) Cold solder. (B) Minimum acceptable wetting angle. (C) Ideal solder.

Fig. 7-8. Quality of soldering.

The areas where the leads of mounted components come through are called *lands.* Holes are drilled or punched through the board at the lands so that components with wire leads can be installed from the other side of the board.

More and more home experimenters are learning to etch their own printed-circuit boards. An experimenter's variation of the printed-circuit board is a uniformly perforated board with copper foil from which unwanted parts of copper can be cut away by hand.

The use of printed circuitry has allowed smaller circuits to be mass-produced quickly and economically. The necessity of wiring between components by hand is eliminated, and, in the case of high production rates, soldering is reduced to two operations. After all the components have been mounted on the printed-circuit board, the copper side of the board is dipped into a liquid flux, and then it is dipped into a solder bath. For low production quantities, hand soldering is used, but wiring between components is eliminated. The marriage between transistors and printed-circuit board has been a happy one because of the small size of transistors and the use of solderable leads on them.

When you are mounting parts on a printed-circuit board, put the component leads through the top of the board and pull the leads through the copper side of the board until the component rests flat against the top of the board (Fig. 7-9A). Then, bend the leads down against the board, along the exposed copper (Fig. 7-9B), and clip them off just beyond the bend, leaving about $\frac{1}{8}$ inch against the land area of the foil (Fig. 7-9C). The small hooks on the leads hold the component in place until it has been soldered.

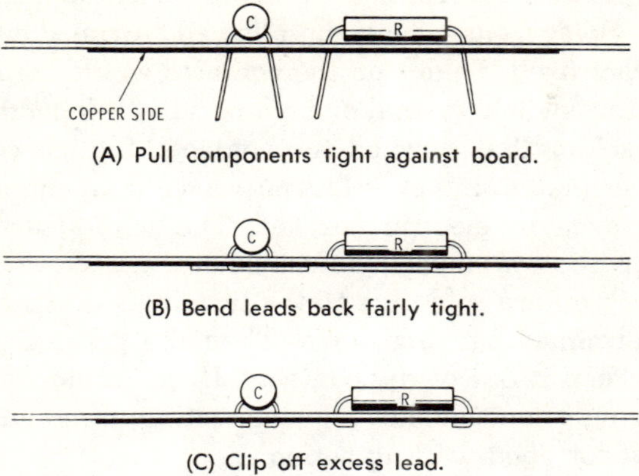

(A) Pull components tight against board.

(B) Bend leads back fairly tight.

(C) Clip off excess lead.

Fig. 7-9. Attaching components to printed-circuit board.

Soldering on circuit boards must be done more precisely than it is on point-to-point wiring. Because circuit boards have small connections, it does not take much time to bring them to soldering temperature, and the heat transfers more quickly, which means the soldering time is less. There must be enough heat for good solder wetting, but not so much that the copper foil is in danger of being lifted from the board. With experience, it should take you only 2 to 3 seconds to solder each connection on a printed-circuit board.

It is most important to apply only a small amount of solder—just enough to fill the crevices between the wire and the land area of the foil (Fig. 7-10). You will have better control of the solder flow if you use a small-diameter solder. Insulated spaces between circuits are frequently very small. You must not allow a bridge of

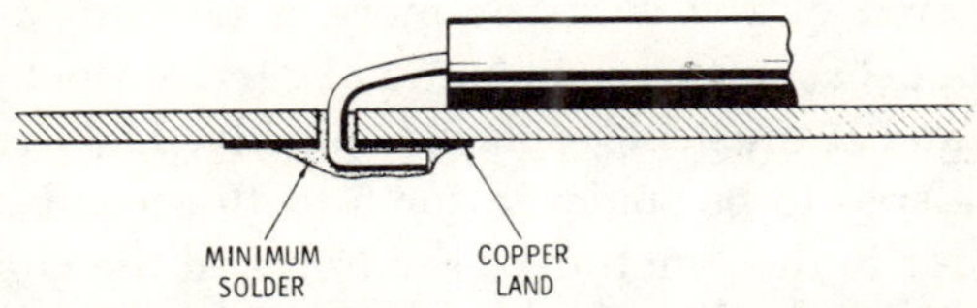

Fig. 7-10. Very little solder is needed for printed-circuit board connection.

solder to form across two circuit sections. If it does, turn the board upside down, apply the iron to the bridge, and let the solder melt back onto the iron tip to open the bridge.

SWITCH TERMINALS

Soldering to switch terminals deserves special mention for only one reason. Solder must never be allowed to run down onto the switch contact itself. Solder on the contacts would impair the operation of the switch or make it inoperative, depending on the amount of solder that reached the contacts. Switch contacts are usually silver plated. Silver is less corrosive than copper, and has lower resistance to electric current. The wiping action of the switch keeps the contacts clean for low-resistance current.

Whenever you are soldering to the terminals of a switch, try to have the terminals hanging down. Then the gravity pull on the solder will keep it out of the contacts. If this is not possible, use solder sparingly and make the connection quickly, but with enough heat for good wetting action.

SOLDERING TO CHASSIS

There are occasions when a direct solder connection to a metal chassis is needed. Most chassis are cadmium-plate steel. The cadmium plating keeps the steel from rusting. The other metal chassis are made of aluminum, which can be soldered to but requires high heat and special solder. Aluminum soldering is discussed at the end of Chapter 2.

Cadmium melts at a low temperature, and soldering directly to the cadmium plating may result in a layer of corrosion that may interfere with good wetting. The cadmium should be scraped away from the spot to be soldered, by using the blade of a knife or the sharp edge of a screwdriver.

If you are going to solder a braided cable to a cadmium-plated chassis, both the chassis area and the braided cable should be pretinned. A hot iron of 100 watts or more is needed to pretin the chassis. This is because the chassis metal carries the applied heat away so fast, and it must be replaced quickly. Hold the flat of a hot iron to the spot to be soldered for 5 to 10 seconds. Apply the end of the solder to the junction of the iron and the chassis. When the solder begins to melt, rub the spot with the iron, firmly applied, and the solder will melt onto the chassis and form a solder coating. Pretin the braided cable as explained previously in this chapter for pretinned stranded wire.

Lay the pretinned wire against the solder-covered area on the chassis and place the hot iron on top, pressing down firmly. The two solder interfaces will melt and run together. Just before you remove the iron, hold the end of a pair of pliers against the wire or braid, and be sure that you do not disturb the wire until the solder has solidified. You will know when the solder has solidified because its appearance will change.

SEMICONDUCTORS

Semiconductors (transistors, diodes, etc.) have indefinite lives, except when subjected to excessive heat, whether developed internally by exceeding their ratings, or applied externally. Excessive heat can destroy them.

Special precautions must be observed when soldering semiconductors into circuits. If soldering is done to the ends of the long leads, as in the case of temporary experimental circuitry, the heat developed by soldering is usually dissipated by the time it reaches the body of the transistor or diode. Any heat that enters the body is usually too little to damage it, unless the iron is held against the terminal for a long time. When a semiconductor is permanently wired into a circuit, the leads should be pulled up short. Under this condition, the heat has less distance to travel up to the body, so something must be done to dissipate the heat. A *heat sink* is used for this. A heat sink is anything metallic that can be temporarily attached to the lead near the body of the semiconductor to carry the heat safely away from the somewhat delicate body of the semiconductor.

You can buy professional heat sinks or improvise your own. The professional heat sink consists of a long, narrow clamp with sufficient metal behind the claws of the clamp portion to dissipate the heat rapidly. If you want to improvise, one of the best and easiest to handle is a solid-copper alligator clip, the kind used on wire leads to make temporary connections. If you don't have an alligator clip, a paper clip will often work, but its broadness prevents the leads from being pulled up short. An effective heat sink is a pair of long-nose pliers. Clamp the end of the pliers to the wire lead, and put a rubber band around the handle to hold the jaws shut. The weight of this arrangement makes it clumsy to handle, but it is effective.

By using a heat sink, you can solder a semiconductor lead to a terminal in the same way that you do the lead from a capacitor or resistor. Be careful not to use too much heat or hold the iron on the connection too long.

To a degree, the same precautions that were applied to soldering a semiconductor apply to soldering the leads of resistors. Excessive heat can alter the value of a resistor, and this is especially important to remember when soldering precision resistors.

CABLE CONNECTORS

The easiest way to solder a wire to a small phone or standard-cord tip is to fill the well of the phone tip with solder, pretin the wire, then apply heat to fuse the two together. Clamp the phone tip upright in a vise. Cut off a small piece of flux-core solder and place it in the well of the phone tip. Apply the flat of the iron tip against the side of the phone tip and hold it there until the solder

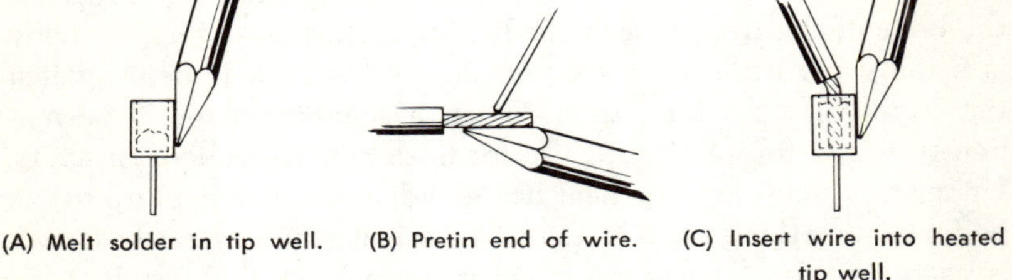

(A) Melt solder in tip well. (B) Pretin end of wire. (C) Insert wire into heated tip well.

Fig. 7-11. Soldering wire to small phone tip.

inside melts. Pretin the end of the wire. Place it in the phone tip well and again apply heat to the side of the phone tip, holding it there until the melted solder results in some give to the wire, and it can be pushed into the well of the phone tip (Fig. 7-11). Let it cool undisturbed.

Multiple-conductor cables may be soldered to multiple-contact connectors in either of two ways. Pretinned leads may be inserted into the pins from the back of the connector, and iron and solder applied to the open end of the pin. Solder will flow into the hollow of the pin, soldering the lead to the inside of the pin. The other method is similar to the phone tip method mentioned above.

Insert a small piece of solder into the hollow of the pin and apply the iron to the outside, melting the solder and pretinning the inside of the pin. Pretin the cable ends and insert them into the pins. Again heat the pins to melt the solder surface together. Care must be taken to be sure no flux flows down the outside of the pin, or solder from the hot iron will wet the outside, and the coating will interfere with the insertion of the plug into its socket. If this happens, scrape the excess away with a pocket knife.

SHIELDED CABLES

Coaxial rf and microphone cables that have outer shields of braided wires are often the most difficult to solder. If you use the procedures explained in the following examples, you should have no difficulty. Cut the outer braid and the insulation between it and

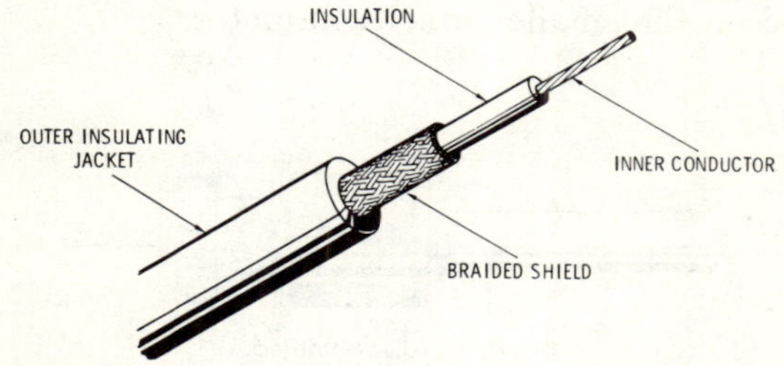

Fig. 7-12. Coaxial or shielded cable prepared for soldering to connector.

the inner conductor, exposing about ½ inch of the bare inner conductor (Fig. 7-12). The inner conductor will be soldered to the center hole or pin of the cable connector. The problem is in making connection between the outer braided shield and the shell of the connector.

On factory production lines, a crimping machine does the best job in the least time. On a one-at-a-time basis, the best method will depend on the type of connector. On microphone connectors, the outer braid is bent back over a spring protector, and a setscrew tightens down on the spring and holds the braid in place (Fig. 7-13). The center wire is soldered to the eyelet in the center of the connector. Coaxial cable connectors vary considerably in their methods of connection. A popular one has holes around a narrowed part of the neck of the outer shell. By pretinning the outer braid and inserting it into the outer shell, solder can be applied

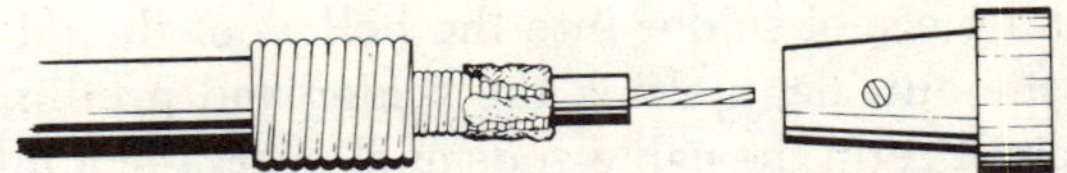

Fig. 7-13. Soldering microphone connector.

into the holes with a hot iron and the two will be joined together (Fig. 7-14). Another method is to twist the braid into a small pigtail and bring it out through one of the holes, then solder it to the sunken part of the shell. A high-wattage iron is required for either of these methods. Be sure to slide the outer screw shell onto the cable before you solder the cable.

When the cable is being inserted into the connector, the inner conductor is, of course, brought through the center pin or hole of the connector. Make the solder connection to the outer braid before soldering the inner conductor. Pull back on the cable a little, then solder the inner conductor. This prevents pull strain from being placed on the smaller inner conductor.

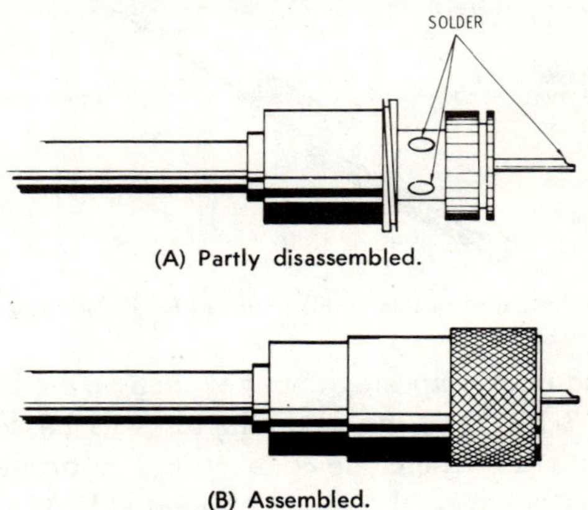

(A) Partly disassembled.

(B) Assembled.

Fig. 7-14. Popular type of coaxial cable connector.

WIRE SPLICES

In electronic circuits, the splicing of wires is normally frowned upon. There should be enough terminal tie points on the chassis to accommodate all components by their own leads without splicing. If splicing becomes necessary, it is important to make a good solder connection and then cover the splice with insulation to prevent shorts to other bare metal. Fig. 7-15 clearly shows how to make the solder connection to a wire splice.

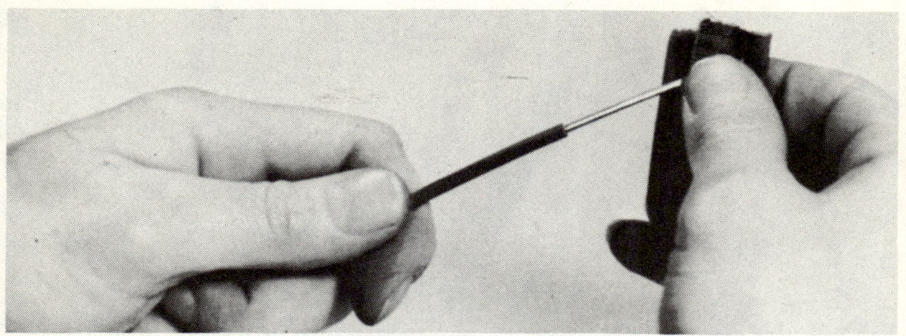

(A) Clean wires with emery cloth.

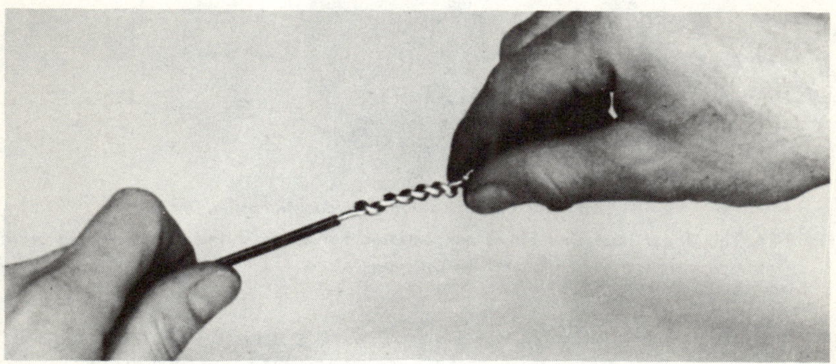

(B) Twist wires together tightly.

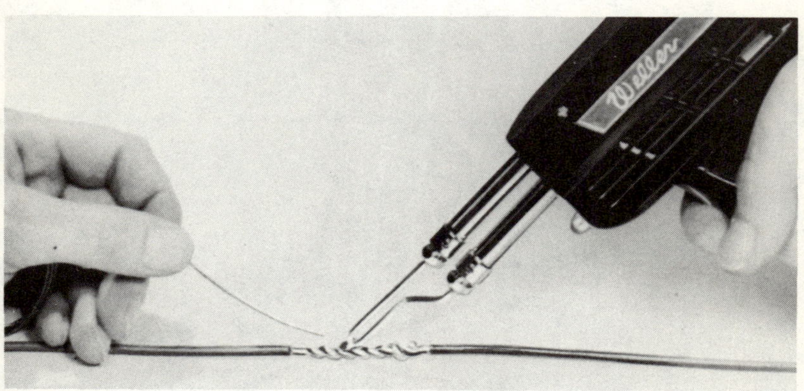

(C) Move hot iron along splice, melting solder.

Courtesy Cooper Industries

Fig. 7-15. Soldering wire splice.

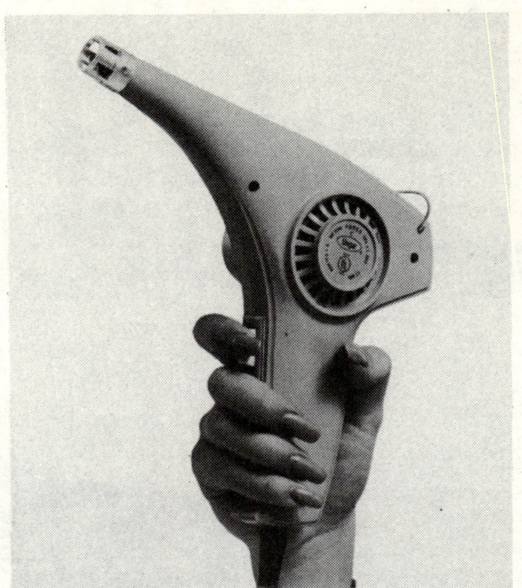

Courtesy Ungar Division of Eldon Industries, Inc.

Fig. 7-16. The Ungar Heat Gun blows concentrated hot air to shrink special tubing used in splicing.

Probably the best way to cover a wire splice with insulation is to use the new heat-shrinkable flexible tubing. This tubing is made of irradiated polyvinyl chloride (pvc for short). Slide a

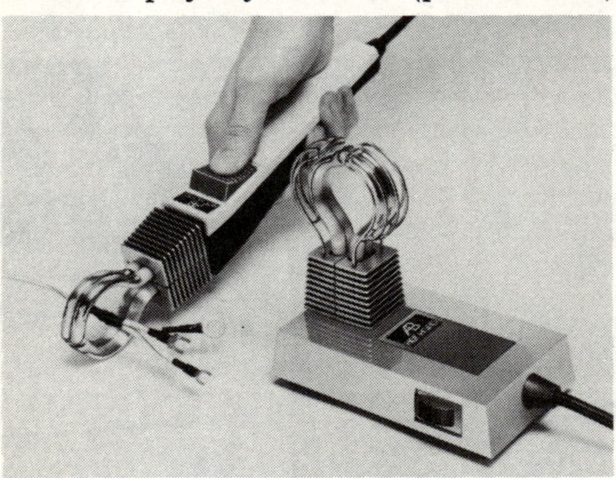

Courtesy American Electrical Heater Co.

Fig. 7-17. The American Beauty Hot Hand closes around a splice, and radiant heat shrinks the tubing.

snug- fitting piece over the splice after it is soldered, then hold a soldering iron or one of the new heating devices near it for a few seconds. The heat will shrink the tubing, making a tight fit that won't slide off.

There are a number of heating devices on the market that are used to shrink tubing over wire splices, and they do it far more uniformly than the side of a soldering iron. Fig. 7-16 shows the Ungar hot air blower, called the "Hot Gun." It develops a very hot and concentrated airstream, from a 3/8-inch nozzle, of 750° to 800°F. Four baffles are included for greater flexibility. Fig. 7-17 shows the American Beauty "Hot Hand," both hand-held and bench models. The heating elements are moved apart so that the splice can be inserted, and then are closed around the work to be heated. Radiant heat is applied, completely encircling the tubing and shrinking it to make a tight fit.

INSPECTION

It is usually pretty easy to see the difference between a good solder connection and a poor one. A good solder connection is one in which there has been good wetting. Good wetting is the result of using a good grade of rosin-core solder on clean terminals and heating the area properly with the correct size of soldering iron.

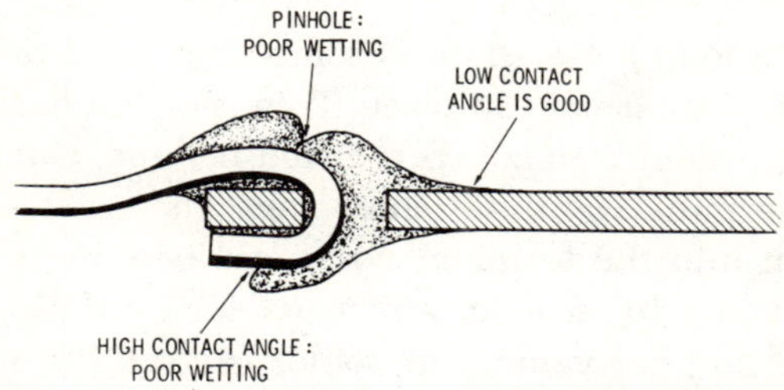

Fig. 7-18. Cross section of wire hooked through a terminal hole.

One way to test an area for cleanness is to put a drop of water on it. A drop of water on a clean surface will spread out; a drop of water on an oily surface will form a small ball. A good solder connection will be bright and will hug the contour of the wires, and the edges will feather out on the wires or terminals with a very low angle between the solder and the base metal (Fig. 7-18).

If the solder balls up and shows a curve at the edges between it and the base metal, it is called a *cold solder connection.* This happens when not enough heat is used on the base metal or it is not clean. It can also happen when solder is applied to the soldering iron and allowed to drop onto the base metal, or roll from the iron onto the metal. In this case, the connection itself has not been heated to solder-melting temperature. There will be no fusion between the solder and the metal. Sometimes the solder of a cold solder connection can be broken off with pliers (Fig. 7-19).

(A) No wetting. (B) Not good enough. (C) Proper wetting.

Fig. 7-19. Poor-to-good solder wetting.

If the solder has a frosty appearance, it is probably because the wires to the connection were disturbed before the solder solidified. The solder is then porous and lacks strength. Fortunately this condition is easily corrected by merely reheating the connection and allowing it to cool undisturbed.

TACKING

In sewing, a long loose stitch is sometimes used to temporarily hold the hem of a garment in place. This is called *basting*. A similar temporary solder connection is called *tacking*. Components are soldered to terminals without shortening the leads, and without hooking them into the terminal holes and bending them around. This makes it easy to unsolder a resistor, for example, and replace it with one of another value. The solder is strong enough to make a temporary electrical connection.

Tacking is easier if the leads of the components and the terminals are pretinned. That is, if solder is first applied to both separately, it is only necessary to apply the iron to the connection to make a tacking connection; no additional solder is needed. Tacking is easier if only one component lead is soldered to a terminal. When several leads are tacked to one terminal, it can be very frustrating to have one fall off while trying to solder another on. So provide yourself with plenty of terminal tie-points in your experimental circuitry.

SOLDERING AIDS

A number of small tools are available that make it easier to get good solder connections. Some are handy for making initial connections; others are for repair (described in greater detail in Chapter 10).

The heat sink mentioned earlier in this chapter is a must when you are soldering to semiconductors. When you ruin a semiconductor with too much heat, not only do you have the cost of another semiconductor but you also have to troubleshoot the circuit again to determine the faulty component. Commercial heat sinks are thin, to allow grasping a lead with minimum lead length. Yet they have good metal bulk for carrying the heat away and keeping it from traveling up the remainder of the lead to reach the semicouductor at the point of entry.

Jeweler's tweezers have fine points that can grasp and hold very small components and fine wires that are too small for needlenose pliers. Some tweezers have offset points for reaching into certain spots more easily (Fig. 7-20). An antiwicking tweezer has a special shape to grasp a wire lead just above the soldering

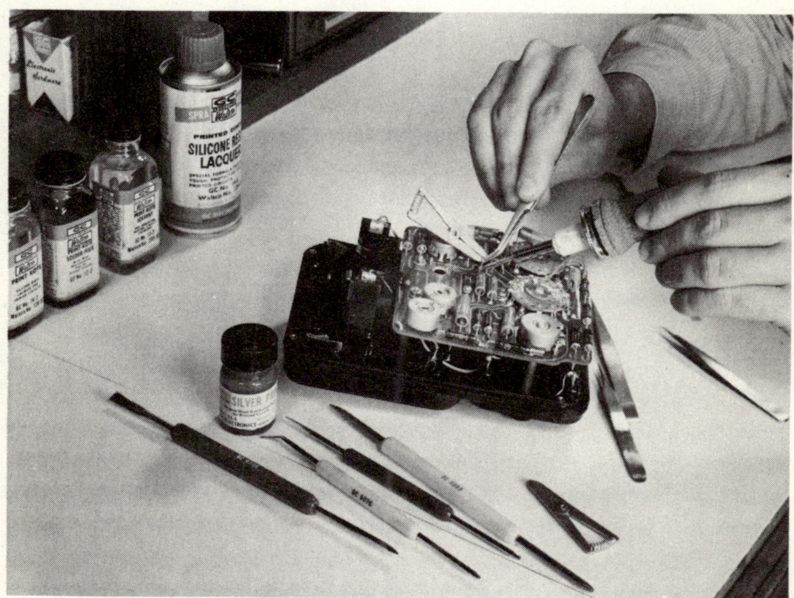

Courtesy GC Electronics Co.

Fig. 7-20. Soldering aids, including heat sink in position.

point. It keeps solder from flowing farther up the lead than is desired.

The double-ended picks, forked tip, stainless steel brush, and scraper shown in Fig. 7-20 are helpful soldering aids. Picks can be used to ream, hold, or pry with; the forked tip is handy for twisting and holding wires; and the steel brush and scraper help to keep the circuit clean while you are soldering.

Hand soldering to a printed-circuit board is easier when the board is held firmly. A special adjustable stand is available for holding boards of almost any width.

Some of the tools shown in Fig. 7-20 are also used for making solder repairs. Solder repairs and desoldering, in particular, with emphasis on the removal of solid-state devices, are described in detail in Chapter 10.

8

Production Soldering

Whether you are soldering a few connections or thousands, the basic requirements for a good solder connection remain the same. The object is to get optimum wetting of the solder to ensure strength and electrical continuity with long life. The methods used to obtain this good solder connection are basically the same also, differing only in the need to obtain a high production rate.

Standard solder connections involving wires and a solder terminal are made with soldering irons. Production needs dictate only slight variations of the hand soldering techniques used by experimenters or technicians. Printed-circuit boards have eliminated the need for hand wiring, but hand soldering is used where the production quantity is low (Fig. 8-1).

From high production rate printed-circuit board soldering, new methods have evolved that eliminate hand soldering. The entire board is soldered at one time by dipping it into a solder bath, and even this has been further refined by wave soldering.

HAND SOLDERING

When hand soldering of standard point-to-point wire connections is called for, one of the standard soldering irons, usually rated at about 100 watts, should be used. In general, when you are soldering one connection after another, a low-power iron, such as that used for experimental or occasional work, would not have the necessary heat reserve. A higher-wattage iron quickly supplies the extra heat needed.

An experienced line solderer will work faster with a hotter iron. Line solderers learn to transfer heat quickly from the iron to the work and then quickly remove the iron. A higher-wattage iron provides the heat needed for this, and experience prevents any burning of insulation or other heat-sensitive components.

For minimum fatigue to the user, standard irons are held with the fingers grasping the handle, with thumb near the line-cord end, and the iron pointing down on the work. This allows maximum visibility of the work (see Fig. 7-5 in Chapter 7).

Courtesy Ungar Division of Eldon Industries, Inc.

Fig. 8-1. Soldering by hand in a plant.

Smaller irons with smaller tips are necessary for smaller connections. When you are soldering connections such as those on printed-circuit boards, the miniature irons with the tiny tips should be used. The connections have little metal mass and, therefore, do not draw heavily on heat from the iron. Irons need not have the large copper core or heavy tip with reserve heat. A typical iron for circuit board soldering is shown in Fig. 8-2. (In Fig. 8-2, note the use of gloves to keep perspiration off the boards, and the fixure for holding the board steady.)

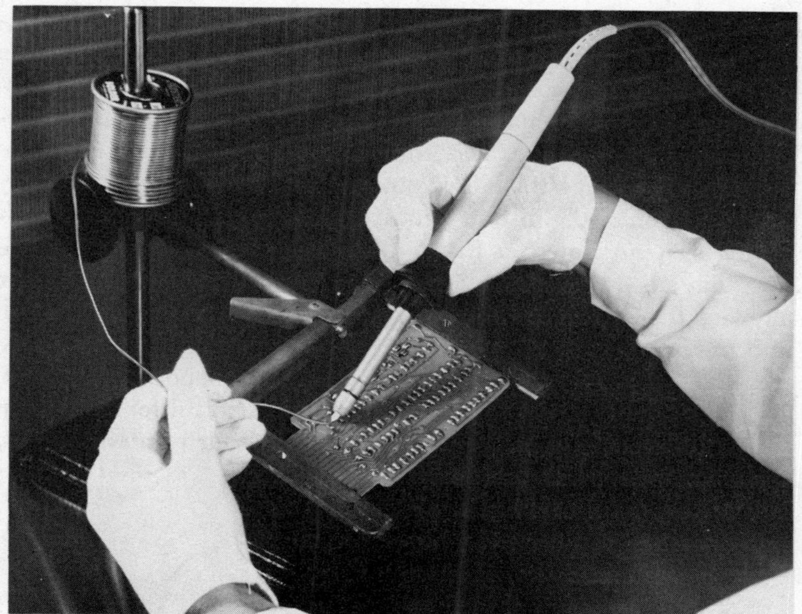

Courtesy Ungar Division of Eldon Industries, Inc.

Fig. 8-2. Imperial iron with conical tip is production tool for printed-circuit boards.

Miniature irons, because of their light weight, are best held like a pencil (whence their name, "pencil iron"). This is shown in Fig. 8-3.

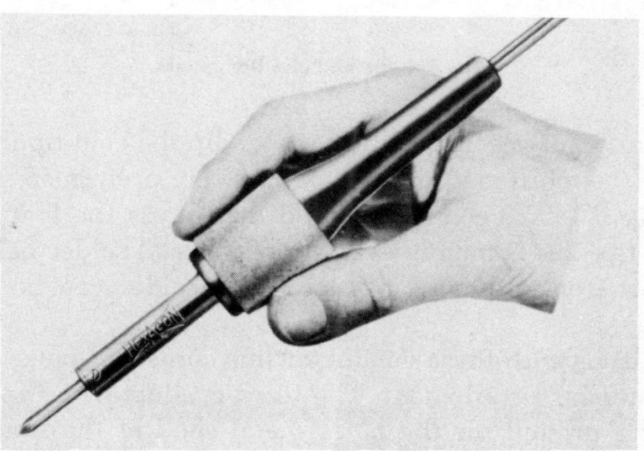

Courtesy Hexacon Electric Co.

Fig. 8-3. Proper way to hold miniature or pencil-type iron.

Instant-heat soldering guns as described in Chapter 5 are, electrically, excellent for production work and could represent quite a saving in electric current when production quantity is low. However, they have one major drawback—they are quite heavy because of their built-in transformer, and would soon tire the arms of even the strongest solderer on the line.

A variation of the instant-heat iron is the Circon *Pulse Dot* soldering system shown in Fig. 8-4. The transformer is a separate unit that sits on the bench and even includes a voltage control for altering the soldering temperature. The iron is light to handle and has a nichrome (instead of copper) loop at the end of the iron that is turned on and off by the foot switch shown at the left. The box contains a transformer and voltage control for setting tip temperature. Fig. 8-5 shows a model designed for printed circuits.

Courtesy Circon Corp.

Fig. 8-4. Circon Pulse Dot system.

The manufacturer's instructions are to dip the cold tip into a flux, then use the moistened tip to pick up a tiny preformed solder ball and apply it to the connection, and then press the foot switch to heat the tip. The company even supplies small solder balls for use with their iron, although standard wire solder can be used also (Fig. 8-6).

Many irons with three-conductor line cords and plugs are available as standard equipment. The third conductor in the line cord provides a ground for the outer metal shell of the iron. Thus, a possible shock hazard to line personnel is eliminated, and one of the hazards in soldering solid-state devices is reduced.

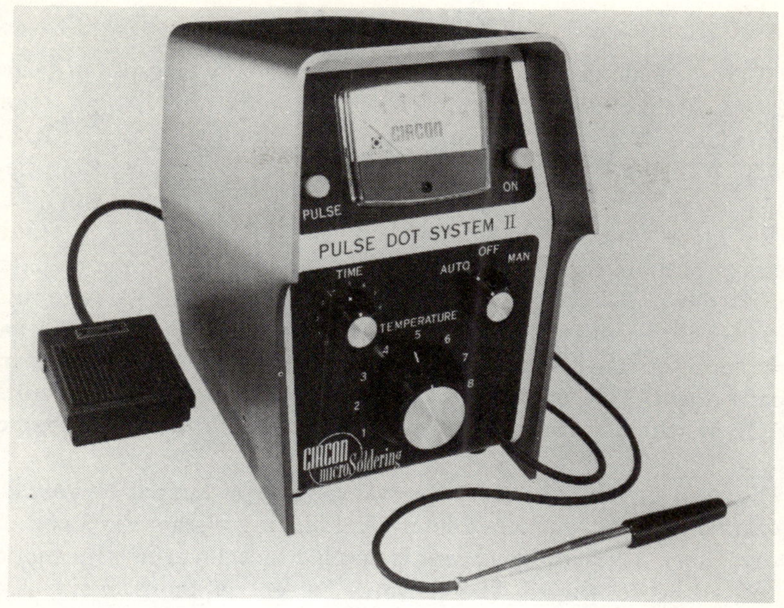

Courtesy Circon Corp.

Fig. 8-5. The Circon Pulse Dot System II has metered temperature control and automatic timing.

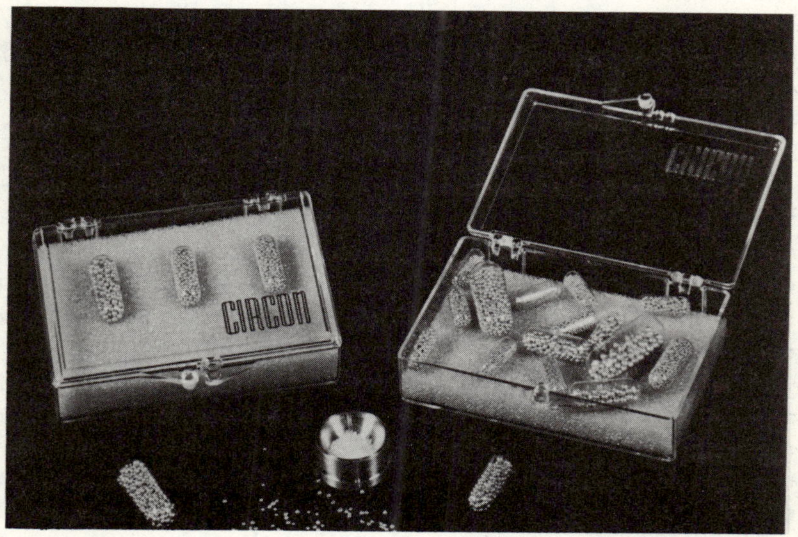

Courtesy Circon Corp.

Fig. 8-6. Preformed solder balls are available in many sizes.

VOLTAGE LEAKAGE

Voltage leakage can be a factor in certain newer types of transistors. Field-effect transistors (FETs) have extremely high input impedance. Some of the newer devices of this type (MOS, MOSFET, IGFET) have a very thin metal-oxide layer stretching between the source and drain and acting as an insulator to the gate. This layer has extremely low leakage and, therefore, a static charge buildup can occur. This charge can become high enough to break down the oxide layer. The transistors are so sensitive in this respect that some are shipped to users with the leads shorted together by a ring, or wrapped in foil, to keep potentials between leads down. This applies also to some of the integrated circuits (IC) in which some of the active components include the oxide layer.

Several possibilities exist for transferring a current or voltage from an iron to a semiconductor. The heating element is a coil of wire. Since alternating current is applied to it, a varying magnetic flux can be developed. The coil may be seen as the primary of a transformer, and any nearby closed electrical circuit as the secondary. If the coil happens to be near a closed circuit including a low-impedance transistor, enough current can be developed in this so-called secondary to burn out the transistor, and this has happened on occasion. The ceramic usually used as the insulator between the heating coil and the metal parts of the iron is a good dielectric for a capacitor. Some ac voltage can be transferred to external circuits by utilizing the capacitive effect of the ceramic as a dielectric, and using the heating coil of the iron as the voltage source. Another possibility is that insulation material is not a perfect insulator. Even if the leakage material has a resistance on the order of several hundred megohms, an external high-impedance circuit of several hundred or more megohms can be exposed to nearly the full voltage of the coil. The last condition, particularly, can be serious in soldering MOSFET transistors.

The solution to the problem mentioned above is to use low-voltage, transformer-coupled irons. The heating elements in the iron are designed to operate from either 12 or 24 volts. This brings the maximum potential well below the breakdown potential of the oxide layer in MOSFET transistors (usually around 100 volts).

The transformer serves to reduce the voltage but, even more important, it isolates the iron from the ac line and transient voltages on the line. One type of transformer-isolated iron is shown in Fig. 8-7, and another is shown in Fig. 4-8 of Chapter 4.

SOLDERING IRON DETAILS

Many soldering irons, particularly those in the miniature class, have handles designed for "clean room" use. Some of these handles are available in a choice of colors.

Tips are made of solid copper or copper with a heavy iron or nickel plating, some are silver plated over the entire shank, and gold plated at the end only. The least expensive tips are solid copper. Copper will corrode at soldering temperatures, and tips of solid copper must be redressed and retinned rather frequently. Iron does not corrode, and iron-clad tips last much longer than the solid copper tips. A silver or gold plating on the iron tip makes it easier to wet it with solder.

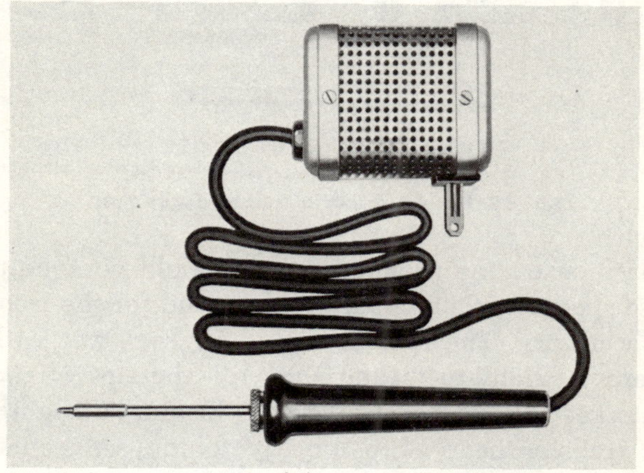

Courtesy American Electrical Heater Co.

Fig. 8-7. Transformer-isolated lightweight pencil iron.

With the high cost of shop maintenance labor, it is more economical to invest in the iron-clad soldering iron tips, even though they are more costly. Iron-clad tips must never be ground down. If they need an occasional cleanup, a few strokes with fine sandpaper is the most that should be done.

Where solid copper tips are preferred, the line foreman should remove them from the irons every night. This will prevent seizure and difficult removal later when they need redressing. When they become corroded, showing dark pit marks, they should be removed from the iron, and the soldering surfaces reground or filed. Finish redressing the tip with sandpaper to take off high spots, and retin it. Start the retinning process with a cold iron. After about one minute, begin to apply flux-core solder to the brightened flat sides of the tip. Within the next minute, the solder will begin to melt. Apply solder to all parts of the reground surfaces.

The ends of copper tips can be ground in shape to better fit the soldering requirements of certain odd contours. Iron-clad tips must never be ground to fit. There is no reason to grind the iron-clad tips because they are available in a wide choice of tip contours (Fig. 8-8).

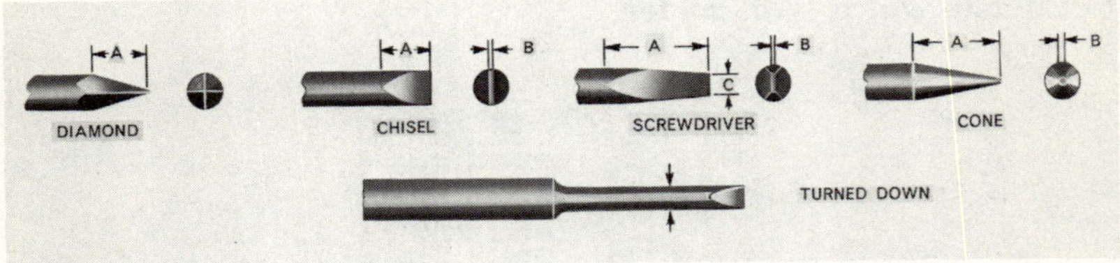

Courtesy American Electrical Heater Co.

Fig. 8-8. Popular shapes of soldering iron tips.

Each position on the production line should be equipped with a cellulose sponge, either as part of the stand for the iron, or as a separate accessory. The sponge should be kept wet with water, and solderers taught to frequently wipe the tips of their irons across it to keep the solder-coated ends bright. A brightly tinned iron will transfer heat more quickly to the connections to be soldered (Fig. 8-9).

RESISTANCE SOLDERING

Applying heat to an electrical connection from the hot tip of a soldering iron is one way to solder. Another way is to pass a high current *through* a connection so that the connection itself can be made hot enough to melt solder (Fig. 8-10). All conductors of electricity have some resistance to current, even though it may be very low. Fig. 8-11A shows a deluxe American Beauty resistance soldering system with adjustable voltage output and adjust-

Courtesy American Electrical Heater Co.

Fig. 8-9. "One-pass" cellulose sponge tip cleaner.

able timing. The two-electrode handpiece is squeezed against the sides of the connection being soldered. The current passing through the connection heats it to solder-melting temperature. Fig. 8-11B shows a low-cost resistance soldering system using an adjustable voltage transformer. The on-off time is controlled by a

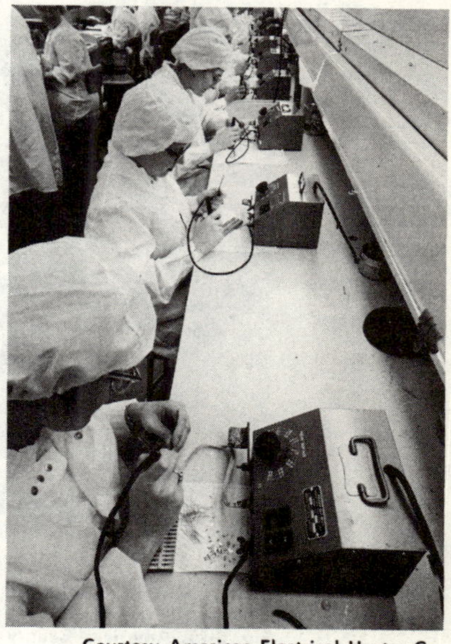

Fig. 8-10. Production-line soldering using resistance soldering.

Courtesy American Electrical Heater Co.

foot switch. Fig. 8-12 shows a two-electrode handpiece for resistance soldering small connections of fixed size.

Heat can also be developed by passing high current through a carbon tip and the metal connection itself. Since carbon has more

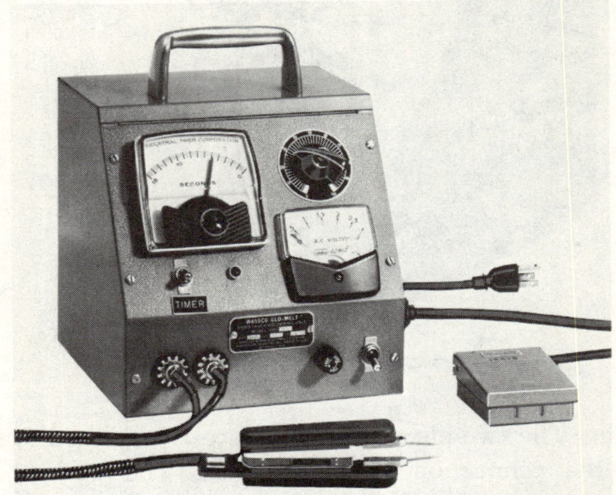

Courtesy American Electrical Heater Co.

(A) With adjustable voltage and timing.

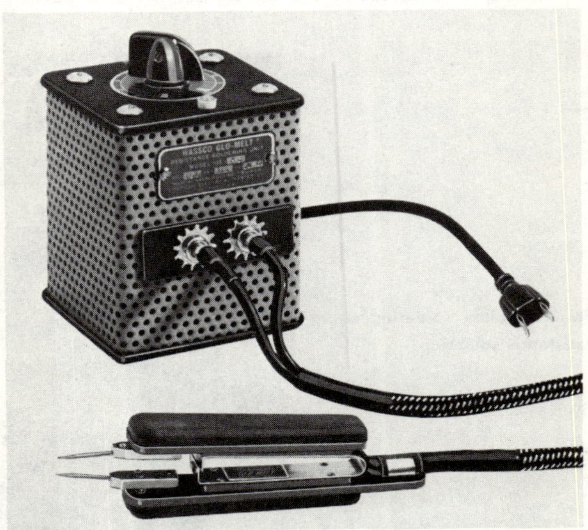

Courtesy American Electrical Heater Co.

(B) With adjustable voltage transformer.

Fig. 8-11. American Beauty resistance soldering systems.

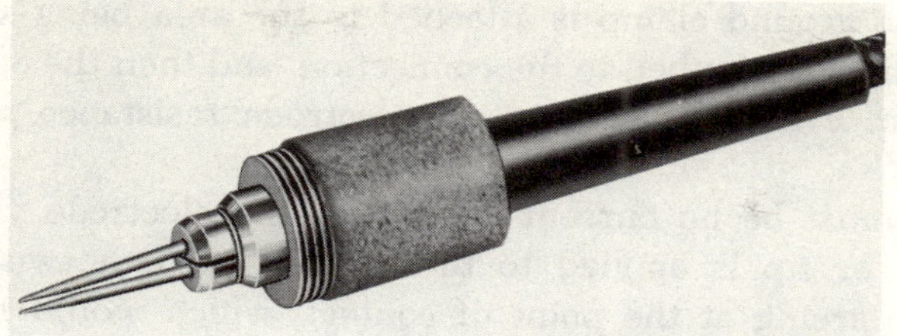

Courtesy American Electrical Heater Co.

Fig. 8-12. Two-electrode hand-held resistance solderer.

resistance than the metal of the connection, it will become hot at the tip where contact is made, transferring this heat to the connection. The circuit is completed through the connection by the use of a grounding clamp on the connection, or the wires to it.

Fig. 8-13A shows a resistance soldering system using a carbon tip and ground clamp. Fig. 8-13B shows how the carbon tip is

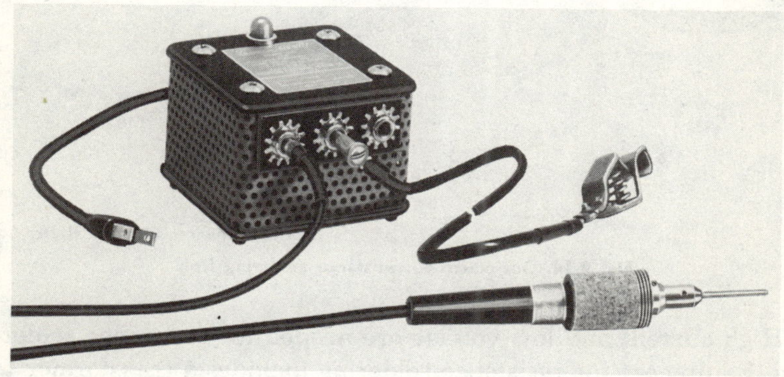

(A) Basic system.

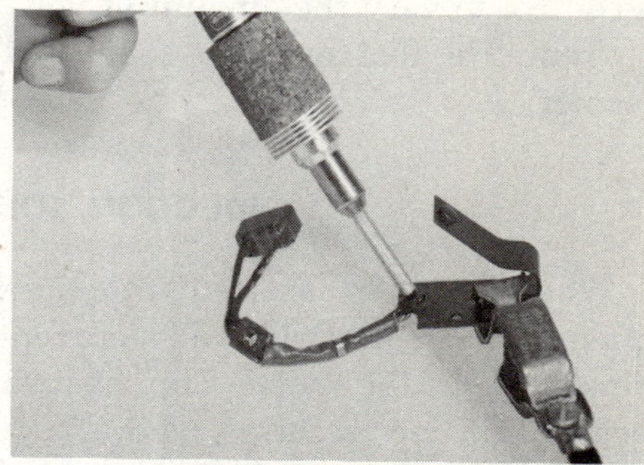

(B) Using carbon tip.

Courtesy American Electrical Heater Co.

Fig. 8-13. American Beauty resistance soldering system with carbon tip and ground.

used: The ground clamp is attached to the area being soldered, the carbon tip is applied to the connection, and then the current is turned on. Fig. 8-14 shows a one-electrode resistance soldering iron.

There must be no current flowing in the electrode when the electrode or tip is applied to the connection. Otherwise, there would be arcing at the point of contact, which would result in corrosion there. A foot switch is used to turn current on after contact is made, and off when the soldering is completed.

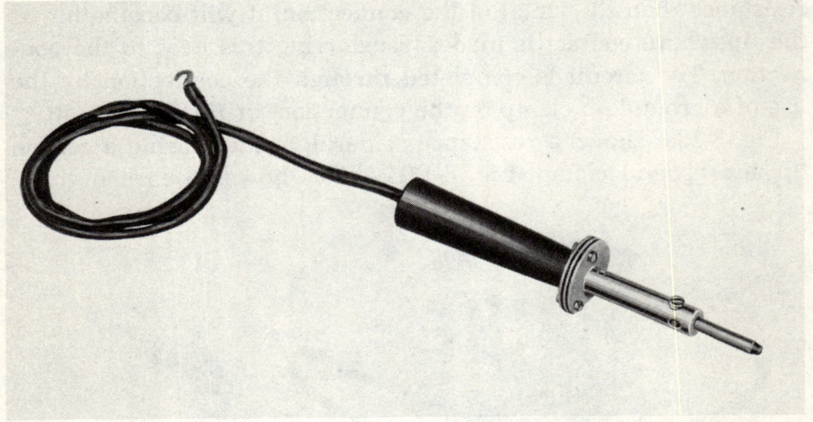

Courtesy American Electrical Heater Co.

Fig. 8-14. One-electrode resistance soldering iron.

High current and low voltage are needed for resistance soldering. Equipment for resistance soldering includes a transformer to step the ac line voltage down. The transformer sometimes includes a control to vary the amount of voltage output, and therefore the current. The foot switch is connected to the primary of the transformer.

INDUCTION SOLDERING

When metal is inserted into the flux of an intense high-frequency magnetic field, eddy current losses result in heat being developed in the metal. This heat melts the solder into the connection. Solder preforms, placed in position before the item enters the magnetic field, are used instead of wire solder. Induction heating is not often used for electrical soldering. It adapts better to the soldering of small individual items rather than electrical connections. When automated, it does result in uniform soldering.

SOLDERING PREFORMS

Certain production soldering jobs are uniform in size and contour. When this is so, it is frequently more economical, and results are more uniform, if precut pieces of solder are used. Specially made, flat pieces or cylinders, called *preforms,* are available from many solder makers. Soldering discs are used, for example, to solder a short wire lead into a metal sleeve as in a feedthrough insulator. Preforms are used exclusively in induction soldering.

THE SOLDERING POT

The soldering pot is a very old device, but it is still used with advantage in electronic equipment manufacturing plants. A sol-

Courtesy American Electrical Heater Co.

(A) American Beauty Pot has a temperature control and holds 1 or 2½ pounds of solder.

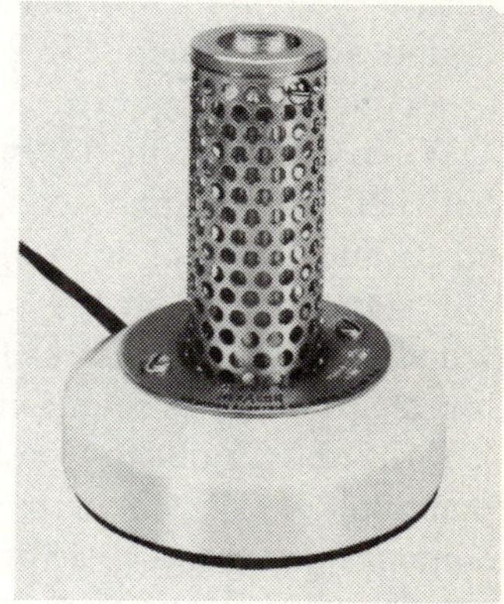

Courtesy Hexacon Electric Co.

(B) One of four Hexacon Mini-Pots available in temperature ranges from 500° to 800°F.

Fig. 8-15. Soldering pots.

dering pot is an electric crucible for heating and holding molten solder (Fig. 8-15). Its main use is for tinning the ends of wire leads.

When wire ends are being tinned, they are first skinned of insulation, then dipped in liquid flux, usually consisting of rosin dissolved in alcohol, and finally dipped in a solder pot for a couple of seconds. The dross that accumulates on the surface of a still pool of solder, as a result of solder oxidation and contaminants from the work being dipped into it, must be frequently cleaned off. The best method is to move the dross to one side and lift it out with a flat sheet of Teflon. No dross accumulates in a system where the solder is constantly circulated (Fig. 8-16).

One of the valuable uses of a soldering pot is for retinning the redressed copper tips of soldering irons. Starting with a cold and dressed copper tip, dip it in liquid flux, then into the molten solder for a few seconds. Lift it out and it will be bright with solder. Add a thicker layer of solder by hand after the tip is put back into use and the iron is brought up to temperature.

DIP SOLDERING

Dip soldering is the general term used for single-operation soldering of the circuit, or foil side, of a printed-circuit board. The bath of solder for dip soldering is similar to the solder pot mentioned above, except for its dimensions. The bath should be held at a constant temperature between 500° and 600°F.

When solder is in the molten state, there will be no separation between the tin and lead, or between the solder alloy and any other added metal such as silver. Zinc and certain added metals will settle out when the solder cools, so the bath must be stirred when first put into use each day. Dross should be skimmed off the top frequently, as mentioned for the solder pot.

Either 63/37 or 60/40 solder should be used in the dip soldering pots. Solder with the eutectic 63/37 ratio is the best because there is no pasty state. A ratio of 60/40 is almost as good because it has a very limited pasty-state temperature range.

When a printed-circuit board is being soldered, it must be held in a frame, secure on all sides, to prevent warpage from the heat of soldering. The board is first dipped in a bath of liquid flux, or flux is brushed on. It is then held at a 2- to 8-degree angle and dipped into the molten solder with a slow sweeping action. It should be in contact with the solder for 3 to 5 seconds, depending on the temperature, size of the board, and the connections on it. The slight angle permits the flux vapors to escape.

Of course, better control of soldering is possible when the action is mechanized. A continuous belt with cams can dip the board into the flux and follow with a floating action across the surface of the solder. This permits the control of temperature versus time for best wetting of the connections on the board. The wavedipper in Fig. 8-16 is used to dip solder pieces with long leads, or pins. It is also a one-at-a-time operation. A high-production method of soldering that uses dip soldering techniques is described in the next section.

The information given above is, of necessity, very brief. Good results in dip soldering require a careful study of requirements and facilities. Suppliers of printed-circuit boards and solder are always willing to give detailed information on dip soldering and the methods best suited for each application. A round-table discussion between the printed-circuit board supplier, the solder and flux supplier, and the customer is the proper start for anyone contemplating dip soldering of printed-circuit boards.

WAVE SOLDERING

Wave soldering is dip soldering in improved form. Where production quantities permit the investment in equipment, wave soldering is today's best method of soldering printed-circuit boards.

High-production wave soldering of printed-circuit boards has two advantages. One, a complete sweep of the board over the solder wave will solder all connections at once. Two, the board speed can be regulated to provide the best control of soldering quality.

A wave soldering system consists of a machine with a conveyor that carries printed-circuit boards through three principal operations—fluxing, preheating, and soldering (Fig. 8-17).

Conveyors are designed to give continuous, automatic operation and can be either parallel or inclined. Parallel conveyors offer the areas to be soldered a lower exposure to the solder wave. Inclined conveyors offer better peelback of the solder as the board leaves the solder wave.

Some conveyors are palletized with fixtures made to carry boards of a specific size, and other conveyors use adjustable fingers that will carry boards of different sizes. A close-up view of a conveyor with fingers is shown in Fig. 8-18.

The first operation through which the conveyor carries the board is the fluxing operation. As with hand soldering, the flux removes the thin film of oxidation from the copper and component leads so that better wetting can take place. The flux may be an activated rosin or a water-soluble flux with an acidic activator. As

(A) Unit with four optional orifices.

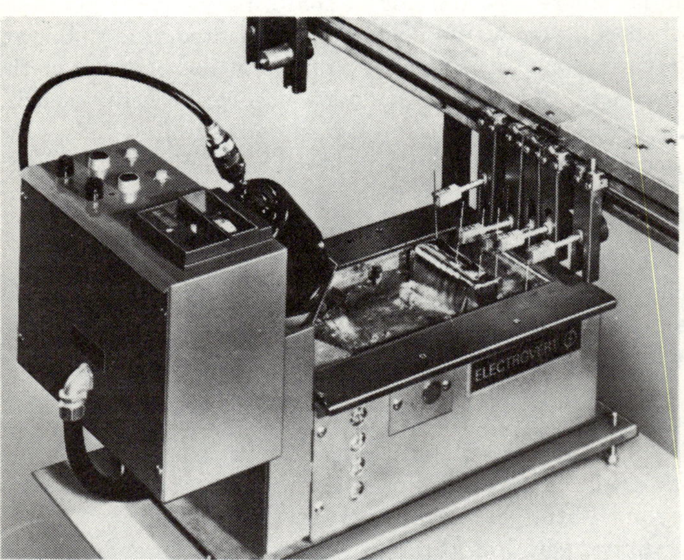

(B) Unit in operation using a rectangular orifice.

Courtesy Electrovert, Inc.

Fig. 8-16. The Electrovert Wavedipper has a pump that circulates the solder and develops a wavelike action in the flow of solder through the orifice.

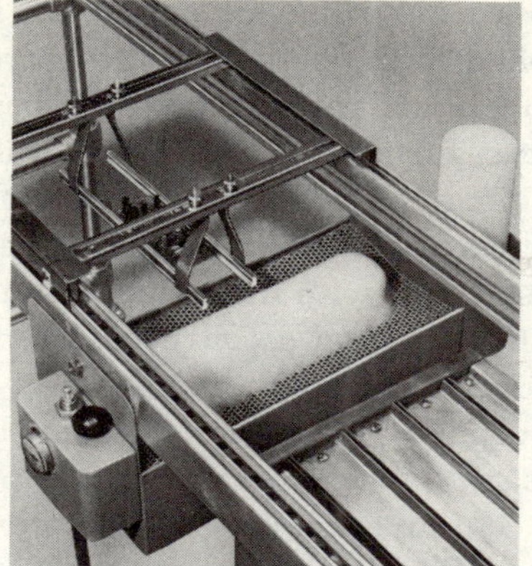

(A) Fluxing of board.

(B) Preheating of board.

(C) Soldering of board.

Courtesy Electrovert, Inc.

Fig. 8-17. Action of wave soldering system.

the board passes over the chimney of the fluxer, flux (usually a foam) is applied to the bottom of the board.

The next operation performed on the board is preheating. Preheating has two functions. One, it evaporates any excess solvent. Two, it conditions the board for the hot solder that follows.

A standing wave of molten solder is created by a pump. As the printed-circuit board with its components installed is passed over the crest of the wave on a conveyor, the molten solder pumped upward through a specially shaped nozzle solders the component leads to the copper of the printed-circuit board; then the excess solder falls back into its container to be pumped again. Because the action is continuous, a cross-sectional view looks like a single wave of solder (Fig. 8-19).

Courtesy Hollis Engineering, Inc.

Fig. 8-18. Soldering machine conveyors may hold boards in two ways: a system of pallets or frames for holding boards all the same size, or adjustable fingers for holding boards of different sizes, as shown here.

There are many solder wave configurations, each designed for a particular soldering job. In general, the wave must be high enough to permit long leads protruding from the bottom of the board to pass through the wave, and the wave must be wide enough to include the full width of the board.

Machines for wave soldering vary in size and complexity. Two examples of several made by Electrovert, Inc., are shown in Fig.

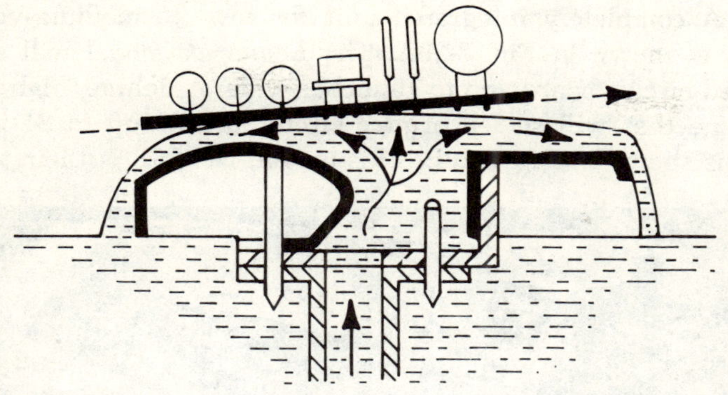

(A) A lambda wave for high-production (18 feet/minute), inclined-conveyor soldering.

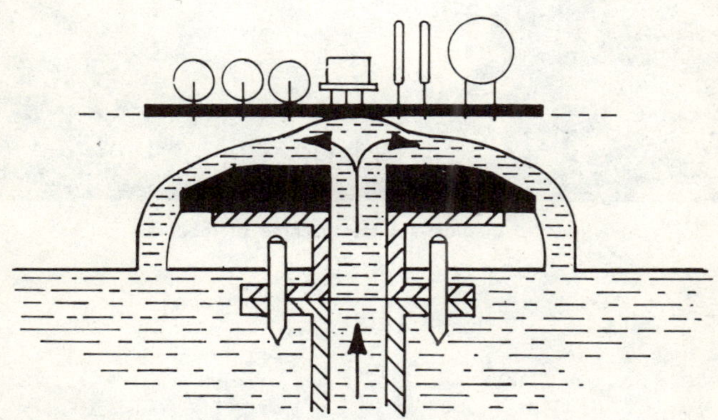

(B) An adjustable wide wave for horizontal systems.

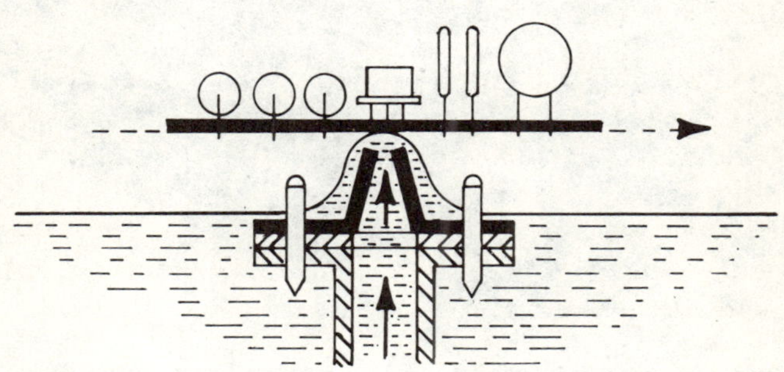

(C) A parabolic wave with a depth of 1½ inches.

Courtesy Electrovert, Inc.

Fig. 8-19. Three types of wave configurations.

8-20. A completely integrated unit for low- to medium-volume needs is shown in Fig. 8-20A. The *Econopak* model will solder printed-circuit boards up to 10 inches wide. A deluxe, high-speed machine that will solder printed-circuit boards up to 24 inches wide is shown in Fig. 8-20B. This model, No. 364, is their top-of-

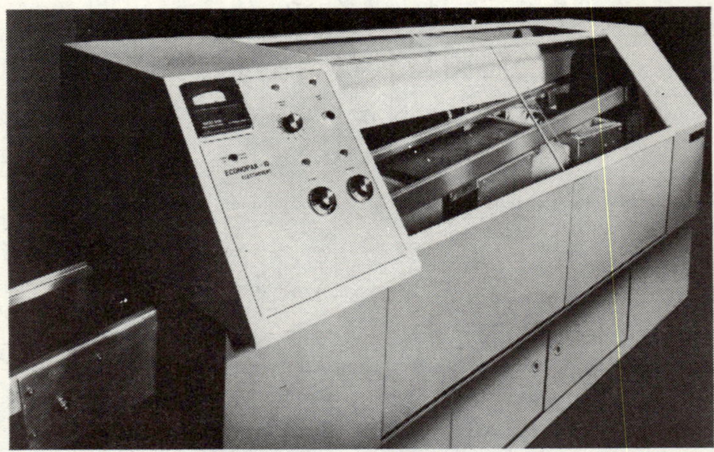

(A) A low- to medium-volume machine of moderate price.

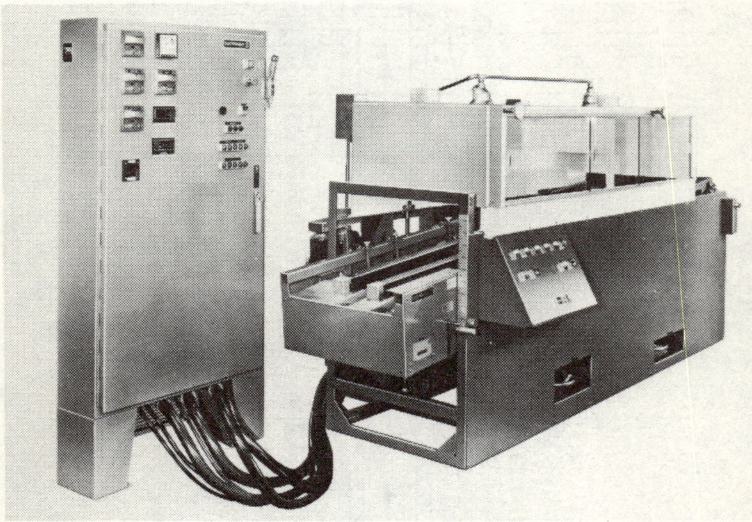

(B) A deluxe, high-speed machine that will accommodate printed-circuit boards up to 24 inches wide.

Courtesy Electrovert, Inc.

Fig. 8-20. Two of many wave soldering machines by Electrovert.

the-line machine and is designed for medium- to high-production soldering.

Another manufacturer of high-production soldering equipment is Hollis Engineering, Inc. In addition to wave solderers, they make lead cutters, cleaners, wire tinners, and assembly lines. They also make specially designed machines (Fig. 8-21).

The preceding is a simplified explanation of wave soldering for high-production soldering operations. A detailed description, in-

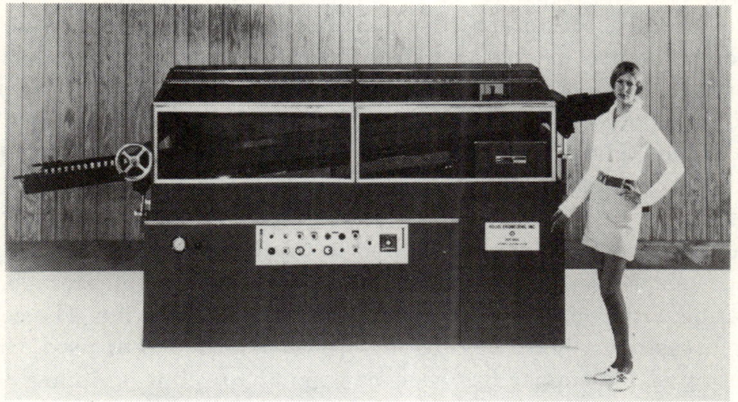

(A) Automatic wave soldering system with finger conveyor.

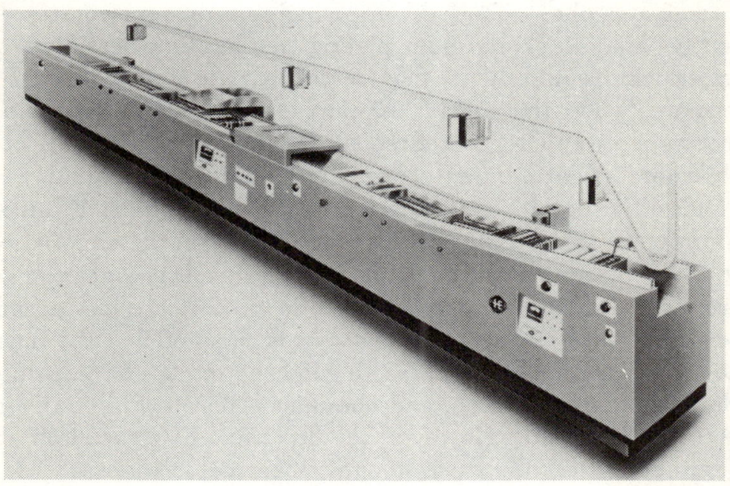

(B) Wave soldering system.

Courtesy Hollis Engineering, Inc.

Fig. 8-21. Wave soldering systems.

cluding the use of tinning oil and cleaning operations, would require a book in itself. If you are interested in installing wave soldering for high production, discuss your needs with one of the manufacturers of wave soldering equipment.

BOARD TREATMENT

One of the big problems in plants doing printed-circuit board assembly and soldering is contamination of the copper surface due, usually, to storage. Everything possible should be done to keep exposure to contaminating conditions at a minimum. The boards should not be wrapped in paper because of the possible effects of the sulfides in the paper. They should not be stored in areas where other production operations are going on, because the air may contain oils or contaminants from the production operations.

Certain coatings can be applied to printed-circuit boards to reduce effects of contamination while the boards are in storage. A coating with certain water-solvent varnishes is effective. The coating is easily washed away just before the board is to be used. Gold plating of the copper has been in popular use, but is finding less favor now. Gold is a nonoxidizing metal, and it dissolves out of the way at the time of soldering. However, the displaced gold soon contaminates the molten solder, which needs replacement frequently. Soldering onto gold plating has more-critical time and temperature requirements than soldering onto bare copper. Furthermore, when the solder becomes contaminated with gold, it becomes more brittle. Of course, if you remove the gold by abrasion before soldering, you prevent the gold contamination.

Solder plating of printed-circuit boards is common. Electrodedeposition of solder puts a thin coat of solder on the exposed copper. The plating must be done on thoroughly clean copper, or corrosion will take place underneath. While the plated solder does not alloy deeply into the copper, the action is completed at the time of soldering. Dissolved rosin flux is a good coating to put on printed-circuit boards for storage, and eliminates the need for fluxing at the time of soldering. However, the flux must be cleaned off after soldering, or the unburned flux may become corrosive.

Printed-circuit boards may be cleaned in several ways. The cleaning may be done manually or automatically as part of a line operation.

Two systems for manually cleaning printed-circuit boards are discussed here. Fig. 8-22 shows several sizes of ultrasonic cleaning tanks. The cleaning solution used in these tanks may be anything from a common household detergent to one of several chemical solutions. The liquid vibrates in the tank at a frequency of about 55 kilohertz, and the tiny bubbles developed penetrate every part of the items being cleaned.

Courtesy American Electrical Heater Co.

Fig. 8-22. American Beauty brand ultrasonic cleaning tanks, in various sizes.

Fig. 8-23 shows a cleaning tank that includes a hand sprayer to clean items in a basket. A chlorinated or fluorinated solvent is sprayed on printed-circuit boards in the basket. The tank is compartmentalized for either hot or cold spraying.

SOLDER RESIST

On some printed-circuit boards with a tight wiring design, the copper circuit foil strips may run very close to each other and it is not desirable to have solder applied to the foil strips. It is frequently preferable to confine the solder to the connection lands only. Since normal dip soldering methods apply solder to all exposed copper, as well as to the connections, it is necessary to put a coat of solder resist on all copper parts of the board, except the connection lands, before soldering this type of board.

Certain organic polymers and epoxies resist solder. A screen design is made to cover all the board except the connection land areas. The chemical is screen-printed onto the copper side of the printed-circuit board. Solder will not adhere to, or creep under, the resist but will be confined only to the untreated land portions.

Courtesy Hollis Engineering, Inc.

Fig. 8-23. A three-stage vapor degreaser uses a hand spray on parts in a basket.

SOLDERABILITY TESTS

A certain amount of corrosion will occur on all exposed copper leads and foils, or even on tinned copper, even if they are exposed to only the oxygen of the air. Corrosion, if it is extensive, will affect solderability. When extensive, corrosion can be cleaned off. But since cleaning is an extra handling, and any extra handling in a factory costs money, it is preferable to determine first if the metals are solderable before any other action is taken.

A solderability test is not difficult, but it must be done by someone familiar with good solder wetting. When you are testing a printed-circuit board for solderability, try hand soldering a few of the exposed lands and then study the wetting action at the edges. Solder should flow out onto the copper with no sharp edge or overlap at the edges. The solder should spread evenly in all directions.

When you are testing wire ends or component leads for solderability, use hand soldering and apply the solder to the lead while the hot iron is on the lead. The solder should melt over the wire and spread out. Gravity will probably pull it down under the wire where it will hang, boat shaped. But the edges of the glob should wet out smoothly on the lead, and the top side of the wire should have a light coat of solder.

A more precise method of testing solderability is to immerse the connection in a heated oil bath. Bring a bowl of polyethylene glycol (Union Carbide's Carbowax 400) or peanut oil up to a

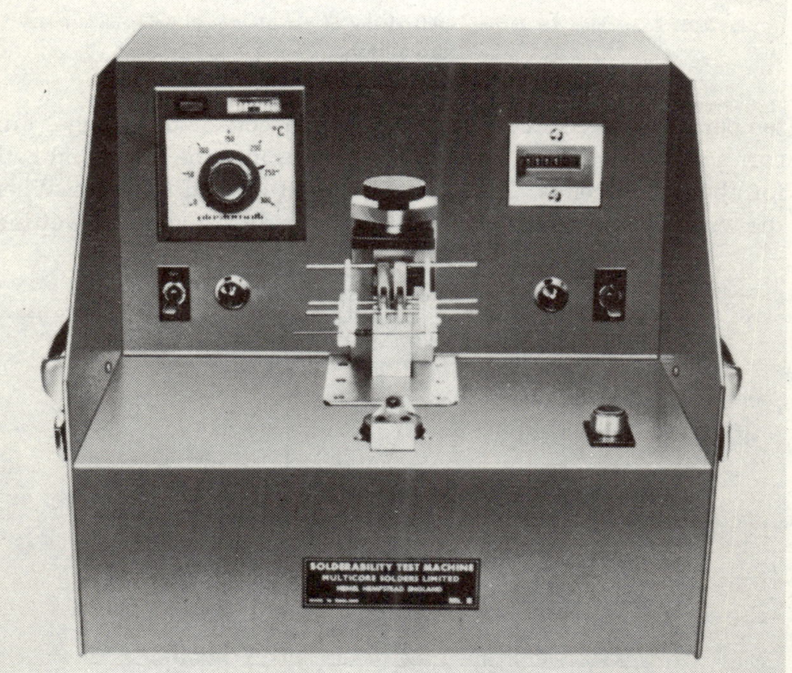

Courtesy Multicore Solders, Ltd.

Fig. 8-24. A Multicore solderability tester.

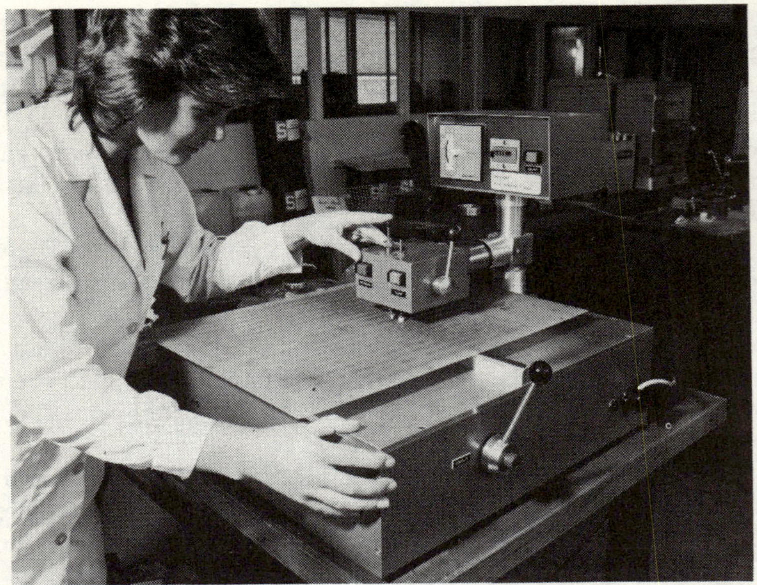

Courtesy Multicore Solders, Ltd.

Fig. 8-25. A machine for testing solderability of plated-through holes on printed-circuit boards.

temperature of 385°F. Wrap a two-turn coil of fine-gauge, flux core solder around the wire or lead. Dip it into the heated oil bath for about 10 seconds. The solder will melt around the wire. When the wire is cooled and the oil wiped off, study the angle of wetting.

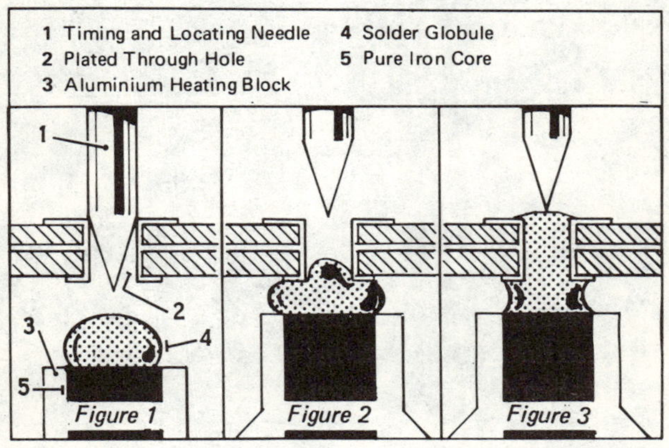

Courtesy Multicore Solders, Ltd.

Fig. 8-26. How the solderability tester of Fig. 8-25 works.

The machine shown in Fig. 8-24 is a Multicore solderability tester, which does somewhat the same thing as the test described above. After the machine is turned on, a piece of the wire or component lead is first fluxed, then lowered horizontally into a globule of molten solder. The machine cuts the solder in two, and the solder flows around the wire and alloys with it. Quoting from Multicore's literature, "The time in seconds for the solder to flow round the wire and unite above it is an inverse measure of the solderability of the wire." This machine is used to test the solderability of wires or component wire leads only.

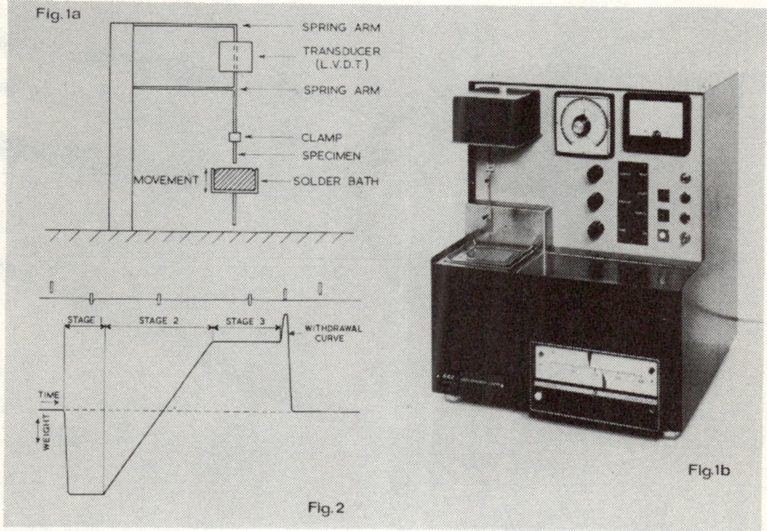

Courtesy Hollis Engineering, Inc.

Fig. 8-27. A solderability tester that permits charting the various stages of solder wetting as a specimen is lowered into a bath of molten solder.

The solderability of printed-circuit boards is often determined by visual examination. However, this becomes difficult for two-sided boards with plated-through holes. The machine shown in Fig. 8-25 performs solderability tests on boards with plated-through holes. The method of testing is shown in the sketch of Fig. 8-26. A pointed rod locates the plated-through hole to be tested, which will be above a molten pellet of solder. The pointed rod is lowered through the hole. The molten pellet and rod are raised. When the solder touches the bottom of the hole, it flows upward, finally touching the point of the rod and making an electrical contact which stops a clock. The time required for the molten pellet to flow up through the hole and make contact with the rod is the test of solderability.

The solderability tester of Fig. 8-27 is based on timing of the solder wetting process. The specimen to be tested is lowered into a solder bath, and each stage is timed and charted as shown. Stage 1 is the time needed to commence wetting; Stage 2 is the rate of wetting; Stage 3 is the fully wetted condition; and Stage 4 is the ultimate wetting as the specimen is withdrawn.

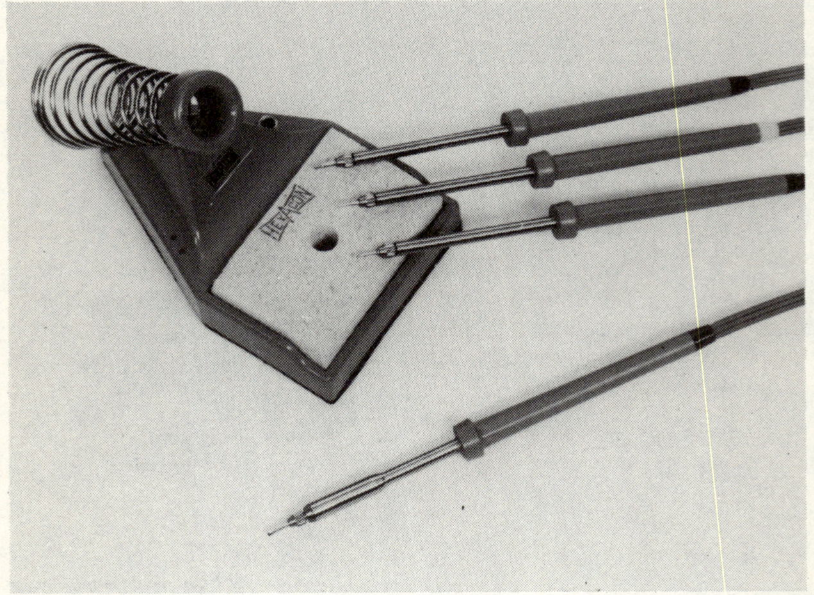

Courtesy Hexacon Electric Co.

Fig. 8-28. Stand for miniature irons includes cellulose sponge to keep the tip clean.

For in-plant soldering instructions, American Electrical Heater Co. of Detroit has produced 25-minute films in full color. These films are loaned without charge to industrial firms.

Firms considering government contracts that include electronic parts for soldering should obtain the NASA *Quality Publication* NPC 200-4 booklet. It is for sale by the Superintendent of Documents, U. S. Government Printing Office, Washington, D. C. 20402. It illustrates the soldering methods required for NASA contracts.

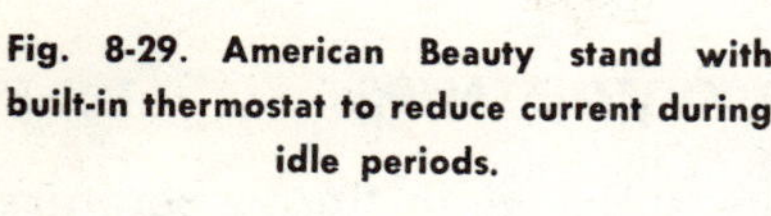

Fig. 8-29. American Beauty stand with built-in thermostat to reduce current during idle periods.

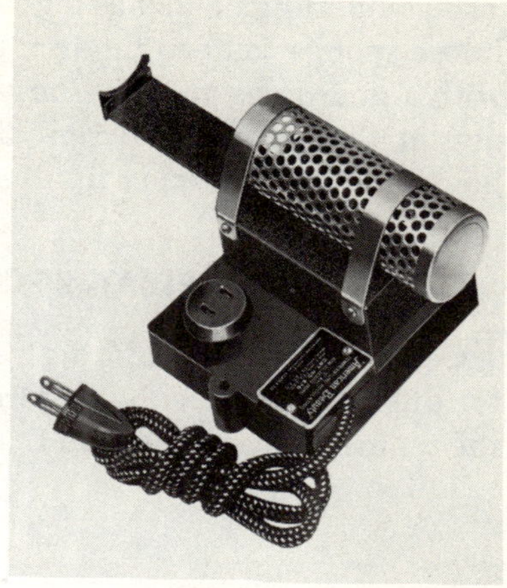

Courtesy American Electrical Heater Co.

STANDS

In the interest of material and personnel safety, all soldering irons should be returned to an appropriate holder or stand during standby periods. Stands may be simple or elaborate, but should hold the iron securely and should either be fastened to the workbench or production-bench or have sufficient weight to stay put.

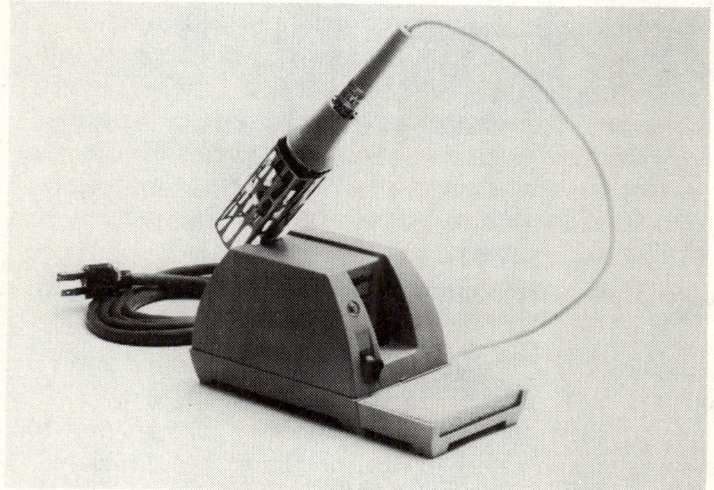

Courtesy Ungar Division of Eldon Industries, Inc.

Fig. 8-30. This stand includes a transformer for reducing voltage to the iron to 24 volts ac. Preset temperature control is a function of tip design.

The stand in Fig. 8-28 has a holder with a helical guard, plus a cellulose sponge to keep the iron cleaned off. Fig. 8-29 shows a stand with a guard designed for large irons. This stand includes a thermostat that reduces current through the iron while it is in the holder, thus prolonging tip life.

TEMPERATURE-CONTROLLED STANDS

Certain types of soldering require a change in tip temperature for different items to be soldered. It is more economical to change the temperature of the iron than to change to a different iron. The

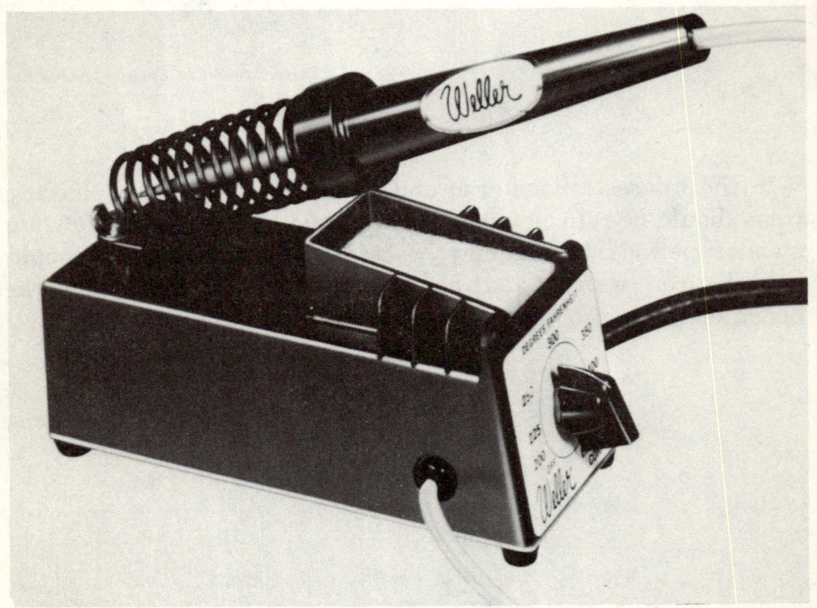

Courtesy Cooper Industries

Fig. 8-31. Weller stand with built-in voltage-control transformer for low-voltage iron.

tip temperature is changed by changing the voltage to the heating element or by adjusting a control on the stand. Many such stands include a transformer, which reduces voltage to the iron and thus gives greater safety to semiconductors being soldered. Many stands are called controlled-soldering stations and come complete with a holder for the iron, a cellulose sponge for cleaning the tip, and an adjustable control for selecting the desired temperature.

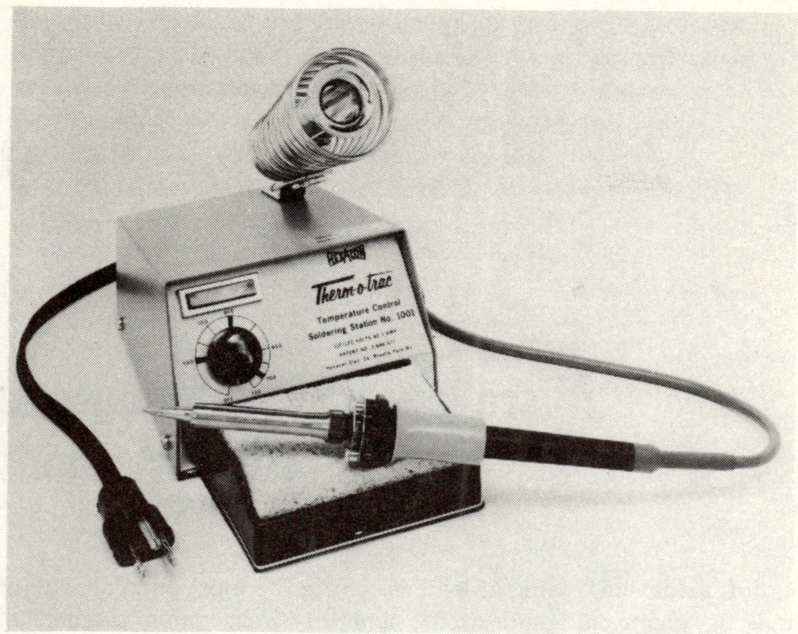

Courtesy Hexacon Electric Co.

Fig. 8-32. Controllable and controlled soldering station. Tip temperature is selectable on a dial. A sensor in the tip feeds temperature information back to a control circuit.

There are a variety of temperature-controlled soldering-iron stands on the market. Three popular brands are shown here. Fig. 8-30 shows an Ungar stand that is available in three preset temperatures. Each stand automatically keeps the tip of the soldering iron at the preset temperature. Fig. 8-31 shows a Weller stand with adjustable temperature. The temperature dial is marked in degrees Fahrenheit. The Hexacon station in Fig. 8-32 has a sensor in the tip of the iron that forms a closed-loop feedback circuit for precise control of any temperature to which the dial is set.

9

Miscellaneous Soldering and Connecting Methods

Soft soldering (with tin-lead alloys) is the most popular way of making electrical connections, and any reasonable method of bringing heat to the connection to melt solder is acceptable. Soldering irons convey heat to the connection. Resistance soldering develops the heat right at the connection. Flame heat may also be used if it is not applied directly to the connection.

THE BOTTLED-GAS TORCH

Gasoline blowtorches have been used for years. When lead pipes were used in plumbing, plumbers used gasoline blowtorches to heat lead for joining the lead pipes. These gasoline blowtorches were too crude to use for electrical soldering, however. The development of bottled propane and butane gases made possible a number of smaller versions of the blowtorch, with smaller and cleaner flames, and these can be used for electrical soldering (Fig. 9-1).

Torch-Heated Tips

One of the best features of the propane and butane torches is a soldering iron tip accessory that manufacturers of all popular brand torches make available. It is attached to the pencil-flame burner and is heated by the flame, from behind (Fig. 9-2). The heat developed is high enough to make large solder connections,

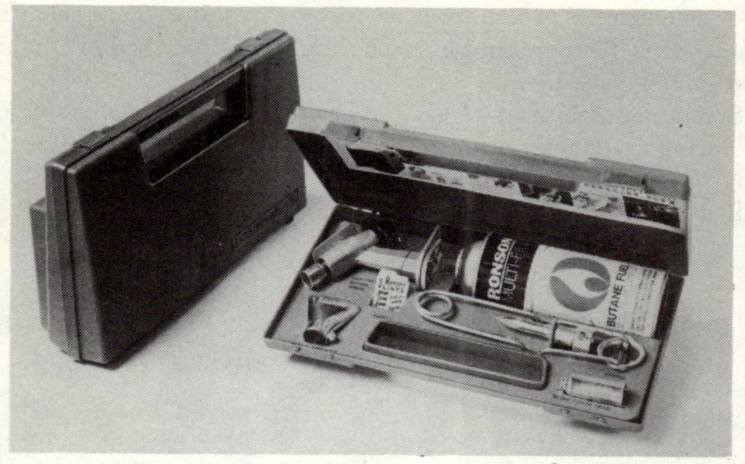

Courtesy Ronson Corp.

(A) A 10-piece butane "Varaflame Torch" kit.

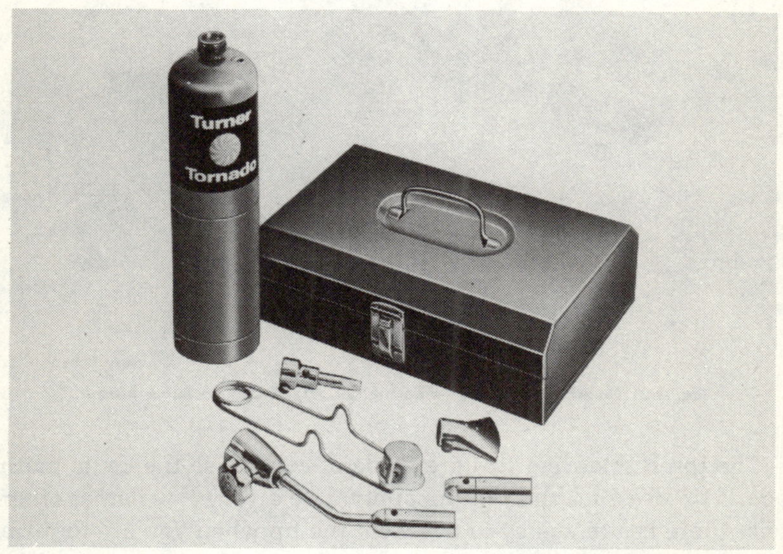

Courtesy Turner Corp.

(B) A 7-piece Tornado torch kit.

Fig. 9-1. Two torch kits used for outdoor soldering.

and is comparable to that developed by a soldering iron with a 100-watt, or higher, rating.

Because of its independence from the ac line and the high temperature it develops, the gas-heated soldering iron is ideal for outdoor soldering of ground rods or heavy wires. If you are soldering small wires, the new cordless soldering irons described in Chapter 6 are more convenient.

Some manufacturers of gas torches make more than one size of soldering tip. However, even the smallest tip is not as small as the tiny tips on most miniature pencil irons. But you would hardly be called on to do printed-circuit board soldering outdoors!

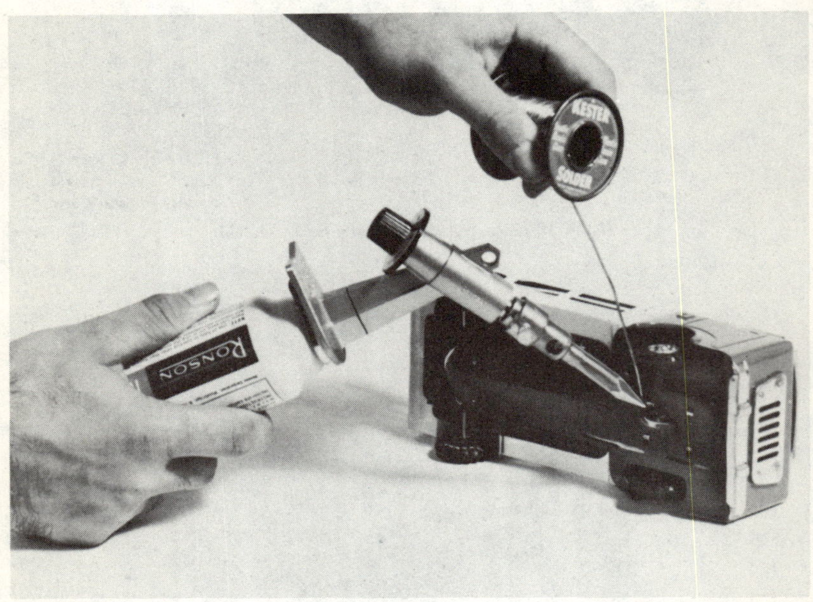

Courtesy Ronson Corp.

Fig. 9-2. Thumbscrew holds soldering tip attachment at flame burner.

The torch soldering tips are made of copper, so the usual methods of tip dressing and tinning apply. Use a very low flame; otherwise there is a tendency to overheat the tip when you are first tinning it or using it. A higher flame may be necessary when working outdoors in a breeze because the movement of air will carry away quite a bit of the heat.

Direct-Flame Soldering

Soldering may be done directly with a torch flame. In fact, it is preferred when the mass to be heated is quite large, such as a ground rod. It is excellent for soldering heavy copper wire to a copper-clad ground rod, or for making solder connections between copper wire and steel, such as steel straps.

When using a torch flame for soldering, apply the flame near the connection, but not directly on it. In the case of a ground rod, for example, apply the flame about 1 inch below the connection. Let the heat travel up the rod to the connection. Apply solder while you are heating the rod, and when the connection is hot enough to first melt the solder, remove the torch. The rod itself, because of its bulk, will retain enough heat to complete the soldering; you will be able to add solder even after the flame is removed. The object is to create enough heat to melt the solder but not enough to char the flux.

It is extremely important that the wire and ground rod, or strap, being soldered be as clean as possible before you start soldering. When a rod is driven into the earth, it is easily contaminated with dirt. Corrosive effects are caused by oils from your hand as it holds the rod in position while it is being driven into the ground. Using emery cloth, scrape the end to be soldered very carefully after it has been driven into the ground. Do the same to the couple of inches at the end of the wire that are to be wrapped around the rod.

Do not be tempted to use the acid flux or acid-core solder that is used with sheet metal work or pipe work. An activated rosin-flux solder is best for electrical connections whether they are made indoors or outdoors. You will be surprised at how easy it is to solder outdoors with rosin-core solder and a torch, once you have learned how. Of course, for the home handyman, and in the shop, the torch is used more for soldering heavy materials rather than for electrical connections. It is an excellent tool for soldering hobby items (Fig. 9-3A), and is probably the best tool for assembling or repairing sheet metal (Fig. 9-3B).

THE FRICTION WHEEL

A novel method of soldering developed by one of the government agencies uses a high-speed grinder loaded with solder. This method makes it possible to solder almost anything: ceramic, glass, and a host of otherwise difficult-to-solder metals such as aluminum, tungsten, stainless steel, and others. No flux is used. The friction supplies the heat and cleans the surface being soldered.

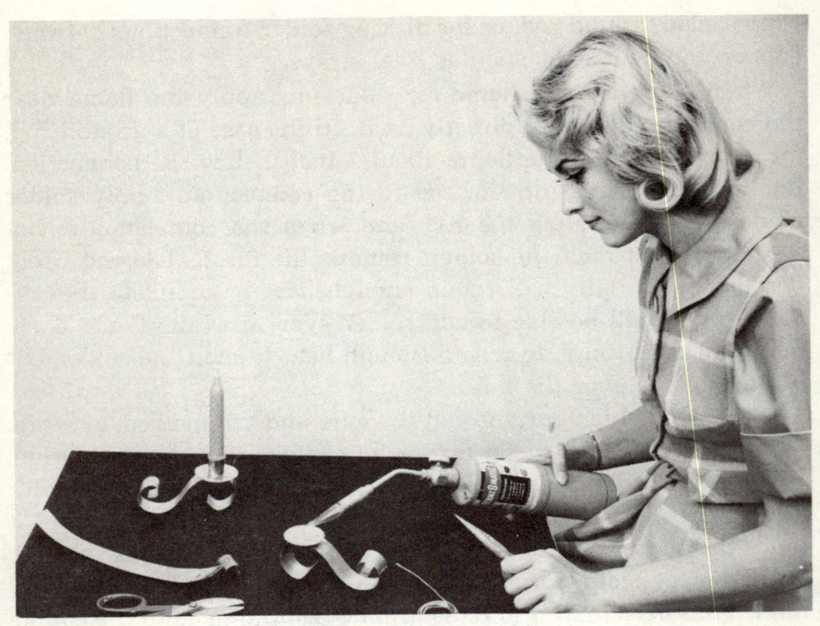

(A) Making copper candleholders.

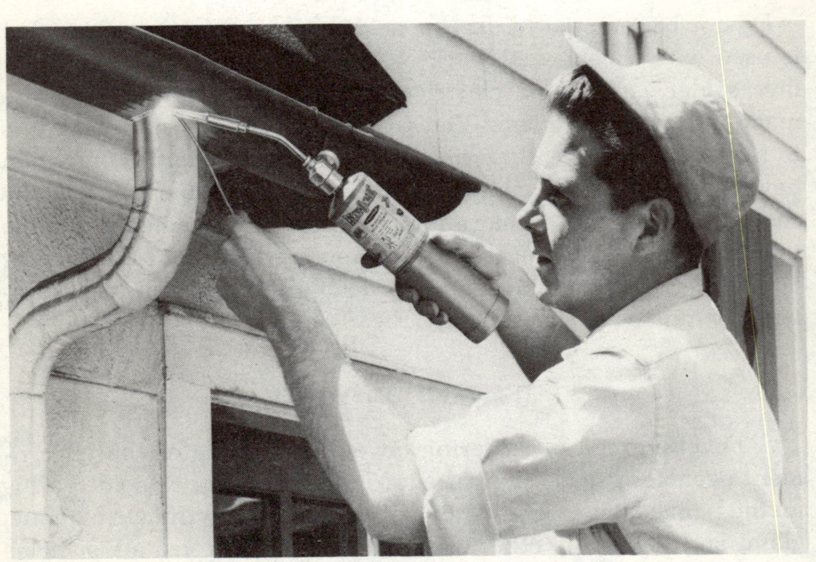

(B) Repairing sheet metal.

Courtesy Bernzomatic Corp.

Fig. 9-3. Uses of bottled-gas torches.

This method uses a high-speed hand drill equipped with a medium-grit grinder. The grinder is heated, then loaded with solder by pressing the revolving grinder against a piece of 63/37 or 60/40 solder without a flux core. The solder is softened to the melting point by the friction of the grinder and will fill the wheel's crevices with solder. Then, with the grinder going at high speed, it is applied to the ceramic, glass, or metal. Again, the heat of friction melts the solder and it becomes affixed to the object being soldered, leaving a thin solder coating on the surface. No flux is

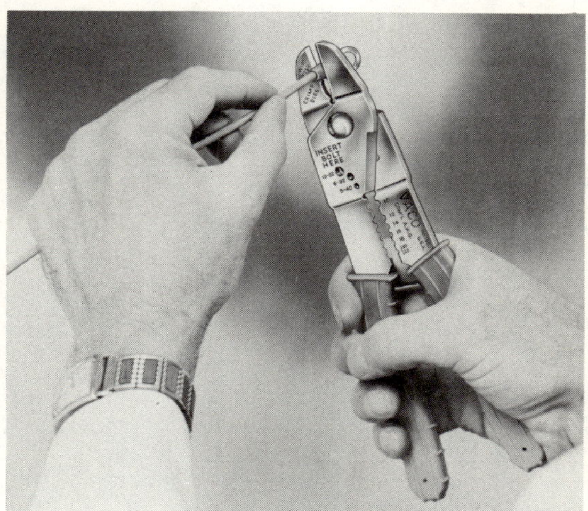

Courtesy Vaco Products Co.

Fig. 9-4. Crimping tool squeezes neck of terminal tight onto end of wire for firm hold.

needed because the high speed of the wheel cleans away any corrosion by abrasion. The solder alloys with the metal before oxidation can take place. In the case of ceramic or glass, there is no oxide involved, of course. Afterwards, electric wire leads may be soldered to the spot with additional solder and a soldering iron.

SOLDERLESS CONNECTIONS

Even NASA and other critical agencies approve the use of crimping to affix connecting terminals to the ends of electrical cables and wires if approved methods are followed. With hollow necks on the terminals, the necks can be squeezed down on the wire to make strong connections (Fig. 9-4). This application applies more to heavy interconnecting cables than to leads on components.

There are many styles of connecting terminal lugs, both bare (Fig. 9-5A) and with insulating sleeves over the wire-holding necks (Fig. 9-5B). They are universally used in electrical connections in automobiles. One handy application is the male and female quick-disconnect pair used so much in cars.

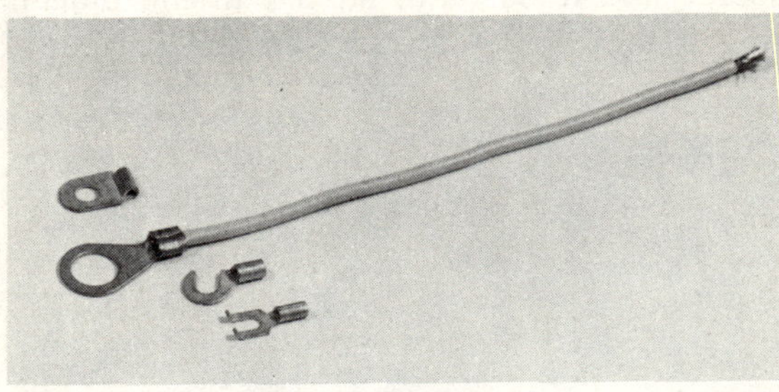

(A) Bare wire terminals.

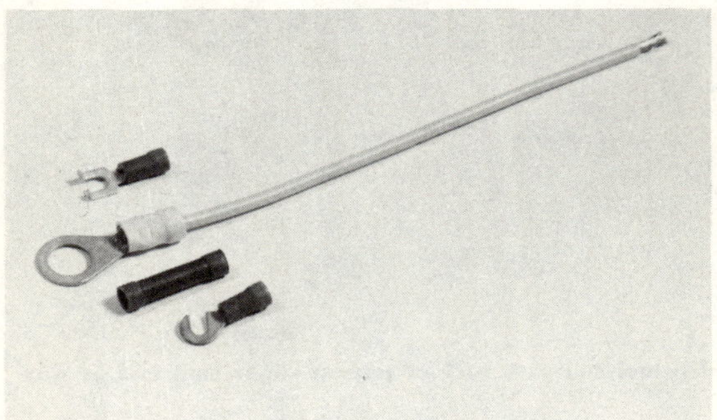

(B) Terminals with insulating sleeves.

Courtesy Vaco Products Co.

Fig. 9-5. Connecting terminal lugs.

The crimping tool illustrated in Fig. 9-6 does a number of things. It cuts wire, strips it of insulation based on wire size, and firmly crimps terminals to wire. (This tool also slices many common sizes of bolts.)

Fig. 9-7 illustrates three methods of splicing wires together using the crimping terminals and the tool. The five basic styles of terminals are shown in Fig. 9-8. These are available in many sizes, nearly all with or without insulation over the neck. Fig. 9-9 shows two styles of quick-connect and -disconnect connectors.

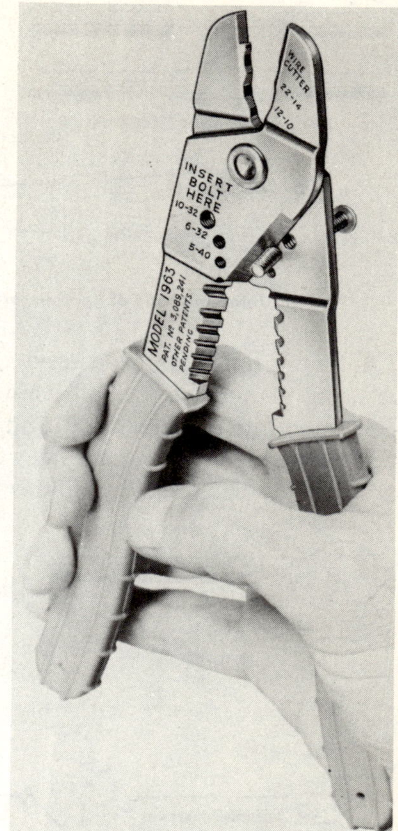

Fig. 9-6. Close-up of a popular crimping and stripping tool.

Courtesy Vaco Products Co.

OTHER METHODS

There are other methods of soldering or making electrical connections, but they are not in general use, and some are still experimental. One obvious method is welding. Small-scale welding is used in the assembly of vacuum-tube elements to their connecting leads. One method based on the discharge of a high-value capacitor is called *percussive welding.* It is used on the tiny leads of microcircuit sections, and eliminates the need for fluxing.

A mechanical connector developed by the telephone company for use on its central dial relay systems is a tight wrap of wire around a fingerlike terminal. A special tool is needed to get the tight wrap.

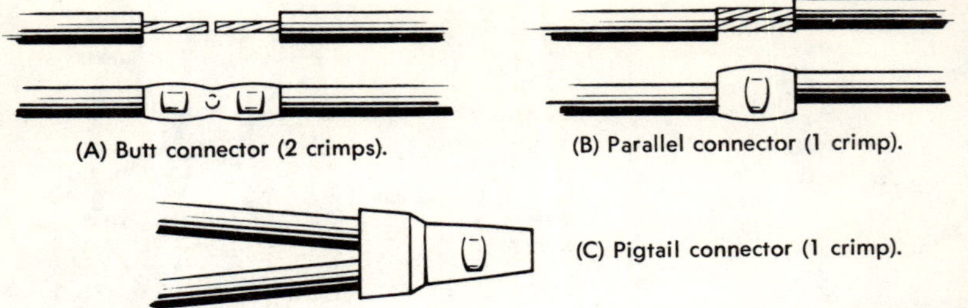
(A) Butt connector (2 crimps).

(B) Parallel connector (1 crimp).

(C) Pigtail connector (1 crimp).

Courtesy Vaco Products Co.

Fig. 9-7. Three methods of splicing wires with crimping tool and connectors.

Infrared light has been used to a limited extent for some time. A beam of light developed from a tungsten halogen lamp is concentrated into a sharp beam, and the resulting heat is used to melt pretinned pieces together. By regulating the width of the beam, a large number of solder connections can be made at one time.

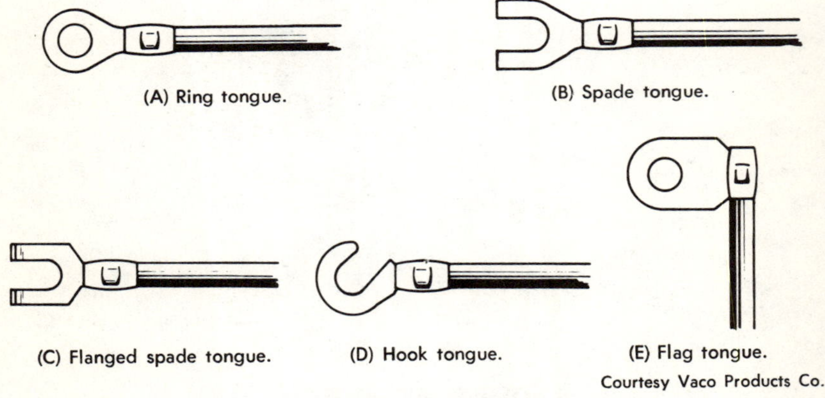
(A) Ring tongue.

(B) Spade tongue.

(C) Flanged spade tongue.

(D) Hook tongue.

(E) Flag tongue.

Courtesy Vaco Products Co.

Fig. 9-8. Five basic terminal styles.

SPECIAL METALS

A number of familiar metals used in electronics are more difficult to solder because of the tenacity of their oxides. Aluminum, for example, oxidizes very easily in air, and a very active flux is required to remove the oxidation. Special proprietary formulas are available from solder and flux makers.

The use of a tin-lead solder is discouraged because of the galvanic action between it and the aluminum. About 1.5 volts is developed between aluminum and a tin-lead solder, which will even-

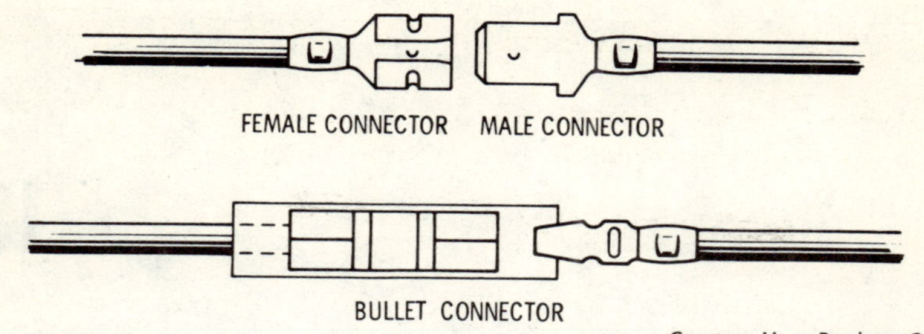

Courtesy Vaco Products Co.

Fig. 9-9. Two styles of quick-connect and -disconnect butt connectors.

tually corrode the connection. Tin-zinc alloys are used to solder aluminum. See Chapter 2 for additional information on aluminum soldering.

A special alloy for soldering silver and for "hard soldering" other metals may be a 90/10 ratio of indium and silver. Hard soldering requires considerably more heat and an organic activated flux.

Leaf spring switches are frequently made of beryllium copper. For a successful soldering job on beryllium copper, a precleaning is often needed. This consists of a wash in diluted sulfuric acid, a clear wash, then a dip in diluted nitric acid, a clear wash, and drying.

10

How to Make Solder Repairs

Anyone who has soldered has had occasion to desolder, or to make repairs or replace parts on soldered connections. The tools you use to solder can be used to make repairs. However, repairs can be done much more easily with special tools designed for the job.

POINT-TO-POINT WIRING REPAIRS

To replace a component soldered to tie-point terminals or tube socket solder lugs, either of two methods may be used. The first method is to cut the lead of the part to be replaced, as close to the terminal as possible. When replacing with a new part, pretin the lead, hold the iron to the terminal until the solder melts, then insert the lead of the new part through the hole of the terminal. The action of insertion should displace molten solder and allow entry unless the terminal hole is already overcrowded. Draw the lead up tight and bring it back over the terminal for at least 180° (or ½ turn). Apply the iron again to the terminal so that the solder in the terminal will flow and fuse with the solder on the pretinned lead. It should not be necessary to add solder, but if the solder already in the terminal and on the lead does not make a good bond, add a small dab of new solder.

The second method is to remove the component and lead. First, find the end of the lead of the component to be removed. Grasp it with long-nose pliers. Heat the terminal with the iron and, when the solder softens, bring the lead around to straighten it. If the end of the old lead is hidden in the solder, the point of a stainless-steel knife blade can pry it loose when the solder is melted. Immediately, while the solder is still molten, pull the component out from the other side. This will leave a hole for inserting the lead of the new component. This operation is made easier by removing some of the solder from the terminal. The best way is with a desoldering iron (described later in this chapter). A standard iron can be used

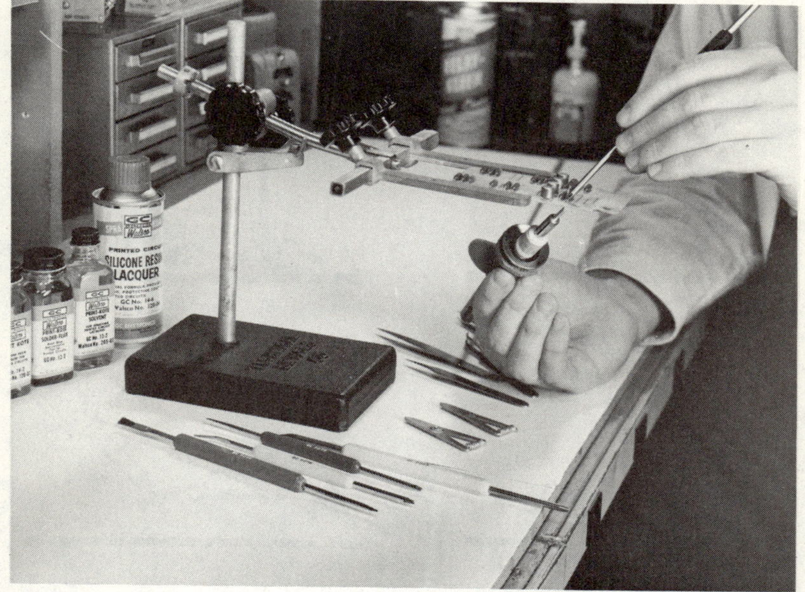

Courtesy GC Electronics Co.

Fig. 10-1. Gravity and capillary attraction draw solder off connection.

to remove excess solder if the work is turned upside down so that the molten solder can slide down the iron (Fig. 10-1).

Avoid excess heat at the terminal when replacing a component. Only use enough heat to soften the solder and then to resolder. An iron with too high a wattage could burn the insulation of a terminal, especially if the terminals are on phenolic-insulated switches. Take care that solder does not run down into the contacts of switches or into the tube-pin contacts of tube sockets. Even a low-wattage iron held too long on a terminal can develop too much heat.

Repairs on printed-circuit boards are a little difficult to make without the use of desoldering aids. One type of repair that can be made without aids is the replacement of a resistor or diode (Fig. 10-2). When replacing one without aids, first crush the body of the resistor or diode by pinching it hard across the middle with a wire cutter. This leaves some of the lead from the component still attached to the board. Bend the leads away from each other at about a 90° angle. Hook the leads of the new component around the old leads. Cut off the excess lead, and resolder.

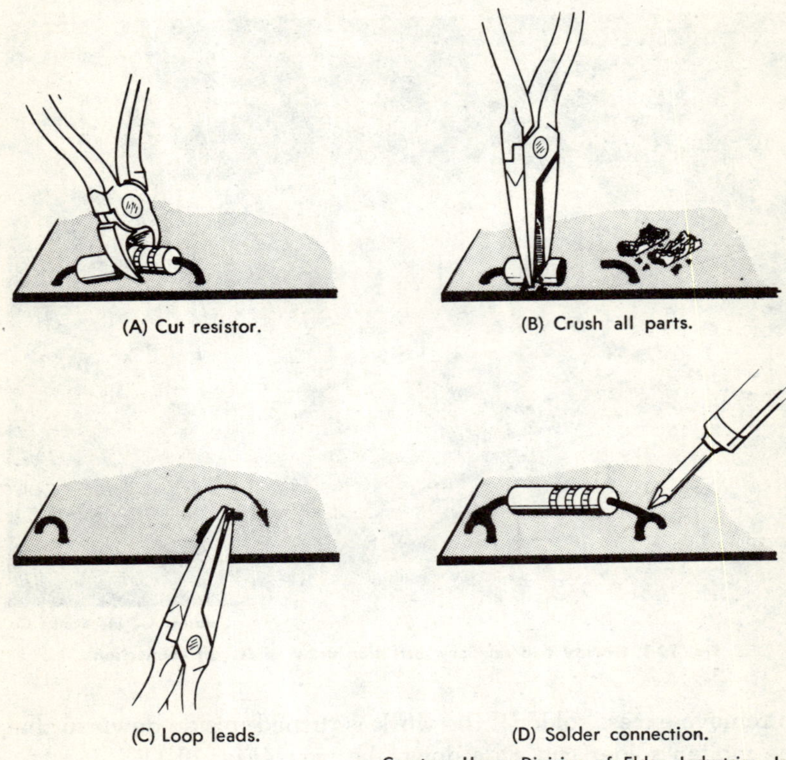

(A) Cut resistor.

(B) Crush all parts.

(C) Loop leads.

(D) Solder connection.

Courtesy Ungar Division of Eldon Industries, Inc.

Fig. 10-2. Replacing resistor on printed-circuit board.

SOLDERING/DESOLDERING AIDS

A number of devices are available to make repair work much easier. Perhaps one of the handiest is the two-ended pick. It is made of stainless steel or of heat-treated, chrome-plated steel. Solder will not stick to the ends. One end of the pick is pointed,

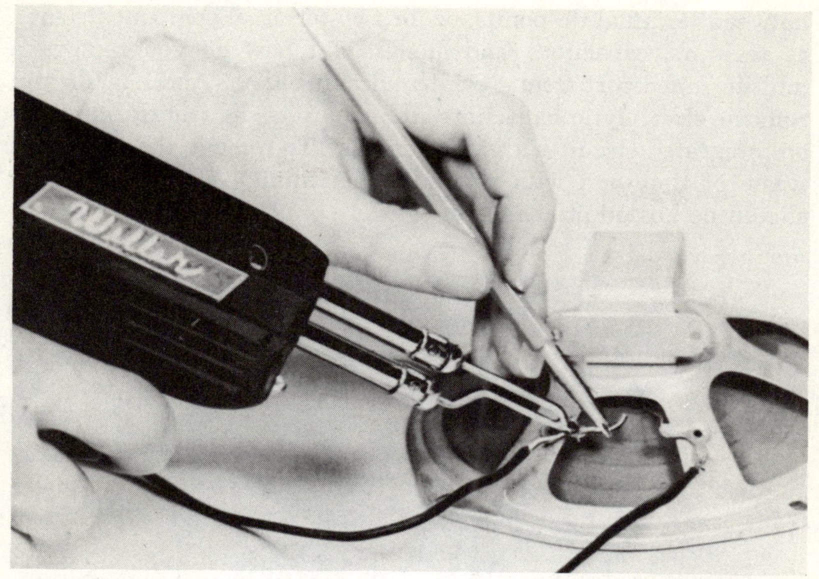

Courtesy Cooper Industries

Fig. 10-3. Forked end is useful for straightening lead for removal.

usually flattened on one side. The other end is bifurcated, or forked, with two prongs. The forked end is ideal for grasping the end of the old lead on the terminal and, while the solder is hot, straightening it out for easy removal (Fig. 10-3). The sharp end is used to poke a new hole through the terminal for easier insertion of the new lead. Some picks have a small brush with stainless steel bristles for sweeping off excess solder. One designed for printed-circuit boards has a brush at one end, and a scraper at the other. With careful use, after the old part is removed it will leave a perfectly clean land area.

Soldering aids were described in Chapter 7 as aids to soldering, but they are also valuable aids for repair work. For example, heat sinks should be used when installing new semiconductors. Tweezers are better than pliers for handling fine wires or tiny components (Fig. 10-4).

SPECIAL DESOLDERING TIPS

The method described in the previous paragraphs is all right for replacing components on point-to-point wiring systems, such as between terminal tie-points, or for two-terminal components such as resistors, capacitors, and diodes. But how do you remove a multiterminal part from a printed-circuit board? And how do you remove electrolytic capacitors, tube sockets, i-f transformers, or an integrated circuit with 16 terminals? To remove these without waste of time and with minimum frustration, all the terminals should be heated at one time.

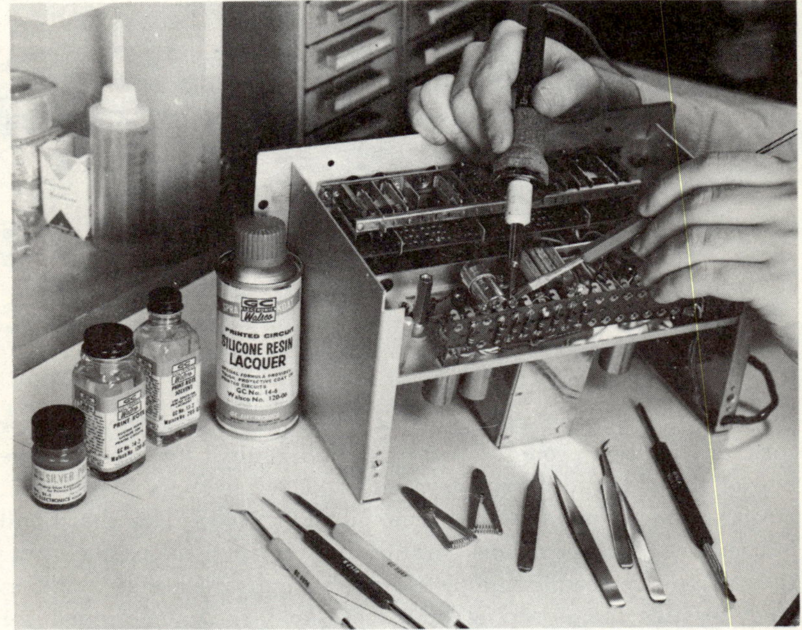

Courtesy GC Electronics Co.

Fig. 10-4. Various soldering aids.

If the tip of your iron were broad enough, it could cover several terminals on a printed-circuit board and desolder them all at one time. Since it is not, the next best thing is a series of tip sizes and shapes for specific terminal schemes.

Fig. 10-5 shows two such tips; one, a cup-shaped tip, is available in four sizes for removing ICs in TO and TO-5 packages. The slotted bar is for use on in-line ICs. The slotted tip heats all the connections of the IC at one time (Fig. 10-6).

A deluxe component desolderer is shown in Fig. 10-7. Not only does this unit include a large number of tip shapes, but the control

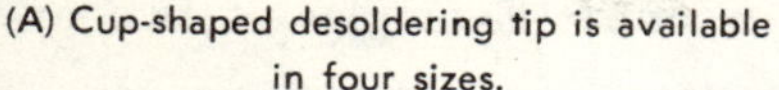

(A) Cup-shaped desoldering tip is available in four sizes.

(B) A slotted bar tip for in-line ICs.

Courtesy Ungar Division of Eldon Industries, Inc.

Fig. 10-5. Desoldering tips screw into various Ungar heating elements capable of developing between 700° and 800°F.

box supplies vacuum for withdrawing molten solder through a hose, and supplies pressurized air for blowing out dirt particles.

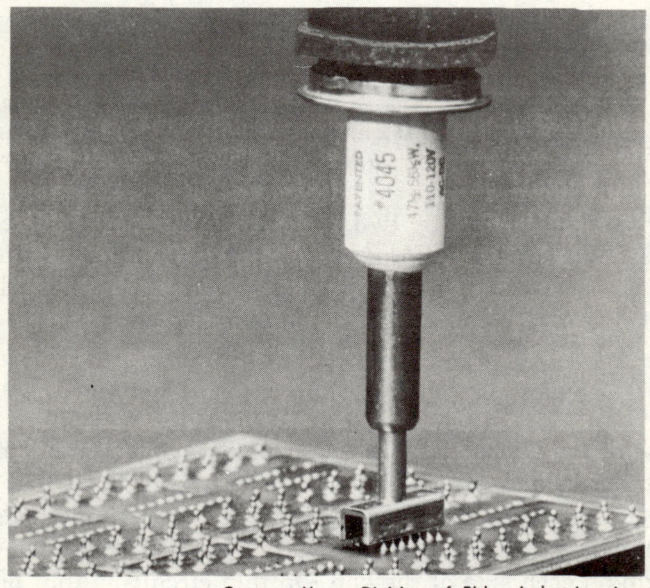

Courtesy Ungar Division of Eldon Industries, Inc.

Fig. 10-6. Slotted bar for desoldering all 16 terminals of integrated-circuit chip at one time.

Some precautions must be observed in using desoldering tips. Don't force the part out of the board; ease it out when the solder on all terminals is melted. The copper foil must be left in good condition to take the replacement component. Be careful not to overheat the land area, or delamination of the copper foil is possible. Keep the desoldering tips well tinned and wipe the tips occasionally on a wet cellulose sponge. All lugs will not be desoldered at once if there are dirty spots on the desoldering tips. Only a brightly tinned tip will conduct heat quickly and efficiently.

To prevent burned fingers and to smoothly remove ICs from printed-circuit boards, use IC extractors. Examples of two ex-

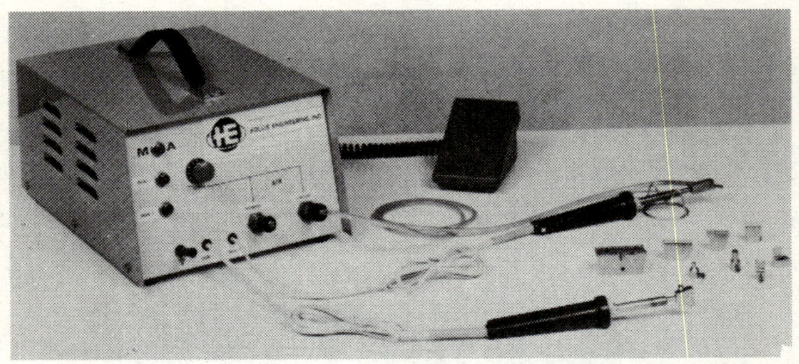

Courtesy Hollis Engineering, Inc.

Fig. 10-7. A desoldering station designed for use in printed-circuit assembly plants. In addition to a wide assortment of tips, the unit includes vacuum and compressed air functions.

tractors are shown in Fig. 10-8. The extractor jaws grip the sides of the IC and exert a uniform pull on the IC from the component side of the printed-circuit board as the terminals are heated from the foil side.

PRINTED-CIRCUIT BOARD WIRING REPAIRS

Some of the copper foil circuit strips on printed-circuit boards are very thin and, if subjected to mechanical abuse, can open. These circuits can be repaired easily by one of three methods. The best method is to bridge the gap with a piece of wire soldered to the strip on each side of the gap. On thin circuits, a thin piece of wire and a small tip on the iron must be used. Frequently, one of the strands pulled out of a piece of stranded wire is about right. Care must be used to prevent solder from bridging across to another circuit.

Another method is to bridge the gap with conductive paint. *Copper Print,* made by GC Electronics, is one such conductive paint (Fig. 10-9). It is applied with a brush (furnished in the cap of the bottle), and it dries rapidly. Here, too, care must be used to

(A) A close-up view of the Ungar Pincess IC extractor that is designed to extract ICs in TO-5 packages.

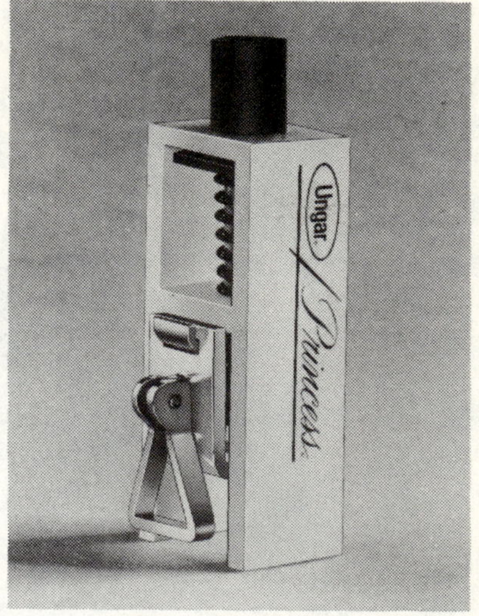

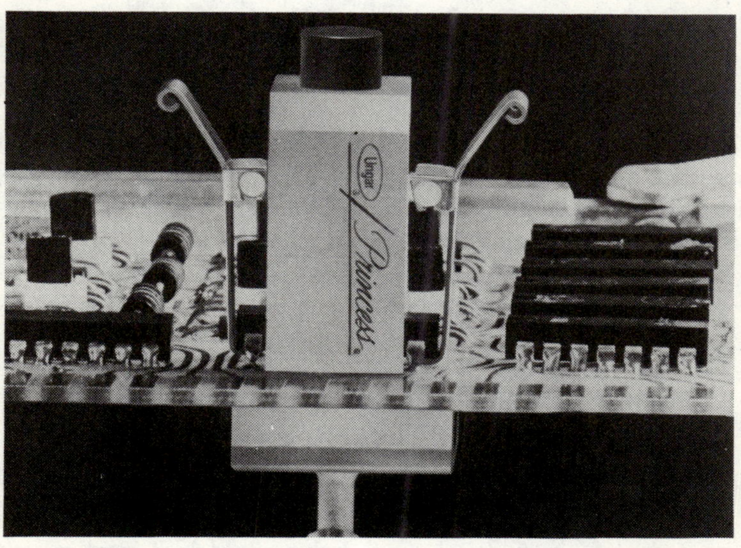

(B) Extractor being used to extract IC from printed-circuit board.

Courtesy Ungar Division of Eldon Industries, Inc.

Fig. 10-8. Two examples of IC extractors.

prevent shorting over to another circuit. While the resistance of this conductive paint is fairly low, it is not as low as a piece of copper wire. It should not be used in circuits where a low impedance must be maintained, such as in a power-supply circuit.

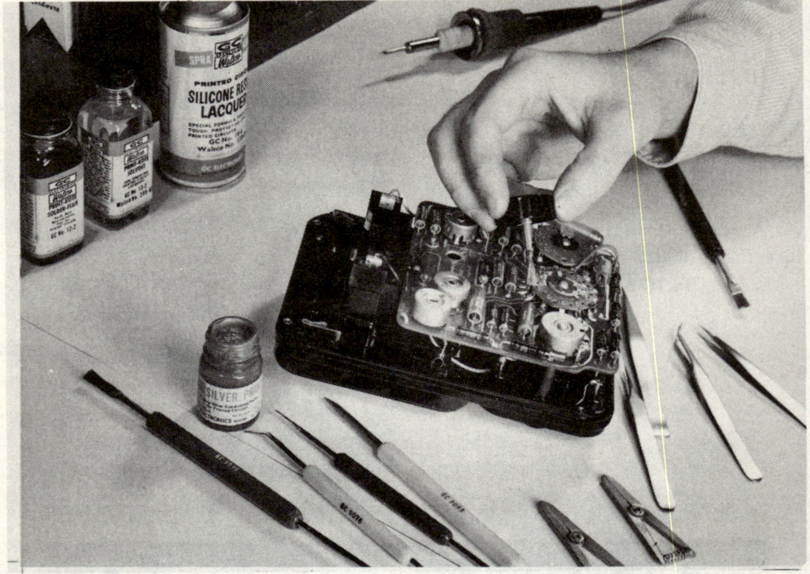

Courtesy GC Electronics Co.

Fig. 10-9. Conductive paints may be used to bridge break in printed circuit.

A third method is to bridge the gap with solder. This is easily done with a forked desoldering tip (Fig. 10-10). Fill the forked tip with solder and, with a little liquid flux applied to both sides of the gap, "write" a bridge of solder across the gap.

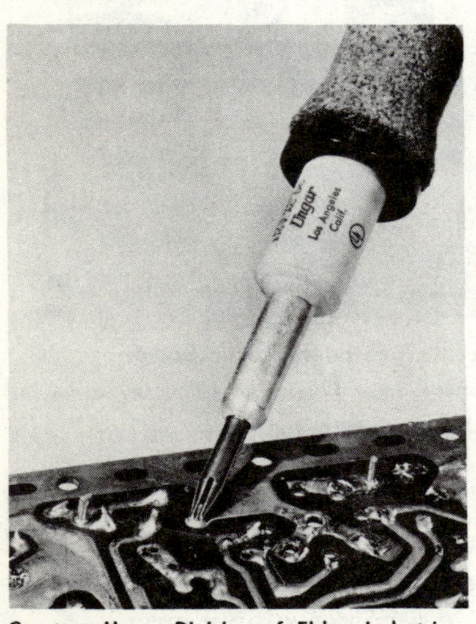

Courtesy Ungar Division of Eldon Industries, Inc.

Fig. 10-10. A forked desoldering tip can be used to flow a solder bridge across a broken printed-circuit board copper connection.

VACUUM DESOLDERING IRONS

Repairs on printed-circuit boards and other terminals are much easier to make if the solder on the connection is removed. While a steel-bristle brush may be used to brush the hot solder away when it is melted by an iron, this has its dangers in that small flecks of solder can get into critical parts of the circuit and cause shorts. Another method of removing solder is to turn the board upside down, and let the solder wick down onto a hot iron. The fastest, the easiest, and the most complete desoldering method is to use a vacuum system that sucks the solder away from the terminal and completely off the board.

Several manufacturers are making desoldering or soldering/desoldering irons that include a vacuum bulb which draws the molten solder up through a tube in the tip of the iron. They all work on air pressure. To use these irons, depress the vacuum bulb and apply the iron to the connection. After the solder becomes molten, release the bulb, and the vacuum draws the solder up through a tiny hole in the tip of the iron into a hollow tube (Fig. 10-11). Lift the iron away and place it over a container or waste area. Depress the vacuum bulb and the solder will be ejected from the tube. The tube should be removed periodically and cleaned. With a little experience, and by using the correct size tip, you can become proficient in desoldering.

WICK SOLDER REMOVERS

One of the most versatile and easily used desoldering aids is a special braided metal with high capillary attraction to molten solder. It acts like a cotton wick in a liquid, thus the name "wick." By placing a small end portion of the wick on a solder connection, and pressing a hot iron against the top of the wick, the solder is drawn up into the wick as the heat of the iron through the wick melts the solder.

Desoldering wick material (Fig. 10-12) is available in various widths. Some manufacturers make as many as five different widths. It is wound on spools and comes in lengths from 2½ to 6 feet. Depending on the manufacturer, widths are as narrow as .025 inch and as wide as .190 inch. The width of the desoldering wick material to use depends on the size of the connection to be desoldered.

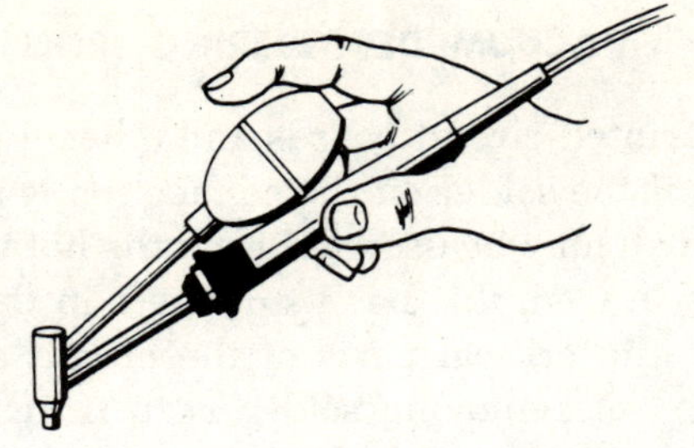

(A) Depressing bulb.

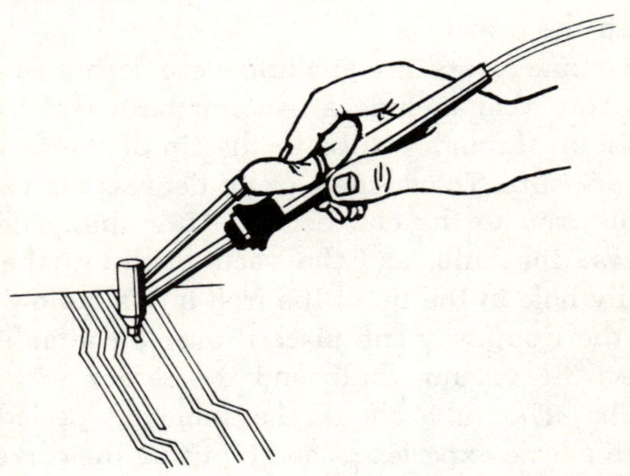

(B) Applying iron to connection.

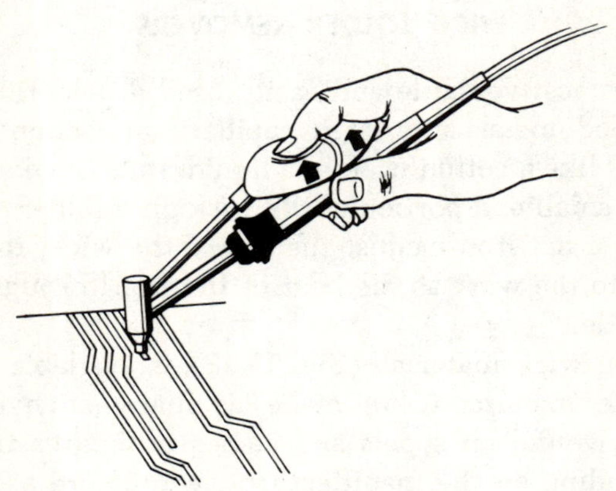

(C) Releasing bulb.

Courtesy Ungar Division of Eldon Industries, Inc.

Fig. 10-11. Desoldering with vacuum system.

For best results, use a chisel-shaped tip on the soldering iron and place the flat side of the tip against the wick. The tip should be slightly wider than the width of the wick material. A brightly tinned tip of this shape will transfer heat quickly. This quick transfer of heat is an important part of wick desoldering. The iron wattage to use depends on the wick width. A 25-watt iron is fine for the very narrow wicks (about $1/16$ inch), and up to 100 watts may be needed for the wider wicks.

Fig. 10-13 illustrates a desoldering wick being used to remove a solder bridge across two lands on a printed-circuit board. In Fig. 10-13A, a wick about the width of the solder bridge is placed on the solder. In Fig. 10-13B, a hot iron is pressed against the wick. In about 2 seconds, heat from the iron will travel through the wick

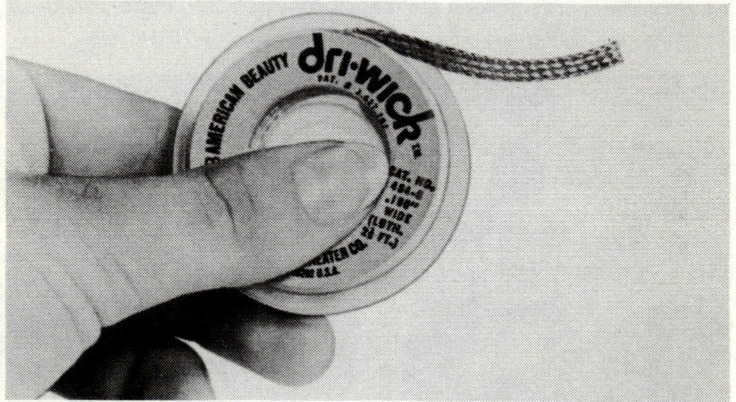

Courtesy American Electrical Heater Co.

Fig. 10-12. The desolder wick is a braided metal with high capillary attraction to melted solder.

to the solder. The solder will melt, and the excess solder will be drawn up into the wick material. As soon as the solder melts, both the iron and the wick are removed. When the end section of the wick becomes loaded with solder, it should be cut off to make new material available for quicker desoldering.

Wick desoldering is quite effective. On the large connections where the component leads are wrapped around the terminal, enough solder is removed so that the leads can be twisted loose with a pair of pliers. On printed-circuit boards with plated-through holes, solder can even be drawn up out of the plated-through holes.

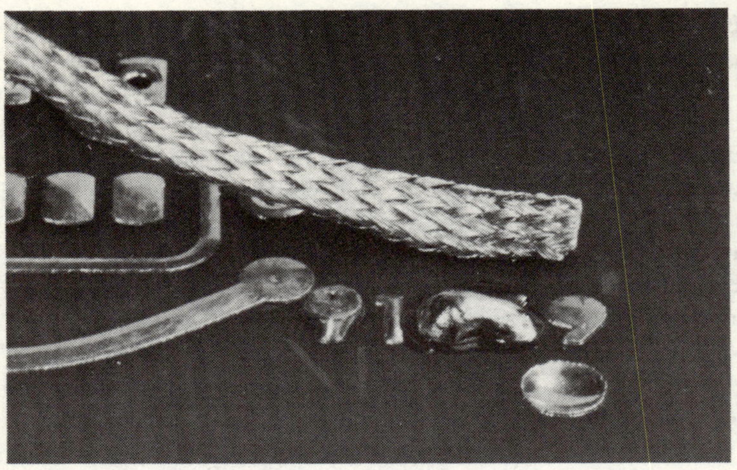

(A) The wick is laid on top of the connection to be desoldered.

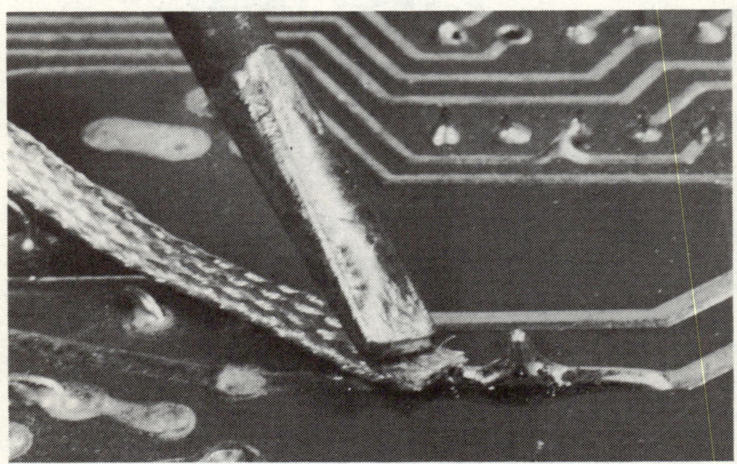

(B) The flat side of a chisel-shaped soldering iron tip is placed on top of the wick.

Courtesy Solder Removal Co.

Fig. 10-13. Heat passes through the "Solder-Wick" and melts the solder which flows up into the wick material.

11

Burns and Their Treatment

The best rule for burns is to avoid them in the first place. A soldering iron develops enough heat to be dangerous to the user or others nearby who might come in contact with the iron. When it is on, it is hot enough to start a fire if it comes in contact with flammable material. For your own safety and the safety of those around you, treat a hot soldering iron with respect. If misused, it can be a dangerous tool.

PRECAUTIONS

One of the first rules for safety is to form good habits for using potentially dangerous equipment. Always lay the iron down in the same place during standby between connections. This helps you form the habit of recognizing that spot as one to avoid. Place your iron in a stand during standby. The safest stands are those with a guard of perforated or wire screen material surrounding the iron (Fig. 11-1). If a stand is not used, be sure the iron tilts back on its handle, the lever point being the heat guard. If the heat guard is not knurled to prevent rolling, screw a broom clip into your workbench, and fasten the iron into it when not using it.

Avoid placing flammable objects on or near the hot iron. This is easier to do when you form the habit of always placing the iron in the same spot. Be constantly aware of people working near you, or standing nearby. Before putting the iron down on the bench or back in the stand, look to see that there is nothing there.

When working with a torch, check to see what is behind the flame. If you are soldering something on the workbench, place a sheet of asbestos or rockbestos material on the bench. If you are soldering something near a wall, tape a cookie sheet to the wall, to prevent scorching it (Fig. 11-2).

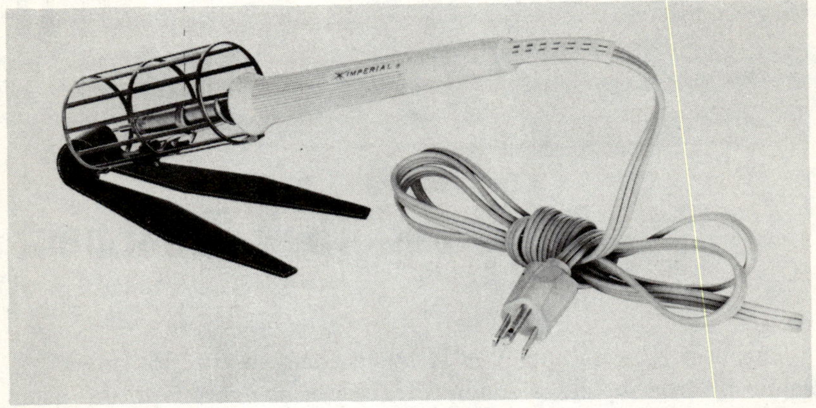

Courtesy Ungar Division of Eldon Industries, Inc.

Fig. 11-1. Inexpensive stand guards hot iron from accidental contact.

Whether through carelessness or sheer accident, burns will occur occasionally when there is a hot iron around. Most burns are small, the result of accidentally touching the iron with a finger, or receiving a flick of hot solder from the tip of the iron. A few burns are more serious, such as those that result from taking hold of the barrel of the iron instead of the handle, or dipping a finger into a pot or pan of molten solder. Spilling an entire dip-soldering pan could be very serious.

CLASSIFICATION OF BURNS

Burns are classified according to the depth of the burn, the area of the body they cover, and the cause of the burn. Causes of burns can be radiation, electrical, chemical, and thermal. For purposes of this book, only burns caused by thermal action from an iron or hot solder will be considered.

Of great concern is the percent of the body that is burned. It becomes important because of the possibility of shock to the body. Shock is possible when 15 to 25 percent or more of the body is burned. However, soldering irons are not likely to cause burns this serious. As an example of a burn area, if the entire palm of the hand and front of all fingers are burned (as when the hot part of an iron is grabbed), this is about 1 percent of the total body area.

The classification of burn depths is divided into three degrees. In first-degree burns, the skin is reddened, sore, and tender, but not blistered. Damage is to the outer layer of skin only. It is not a serious burn, and whether treated or not, it will heal within six days and leave no scars.

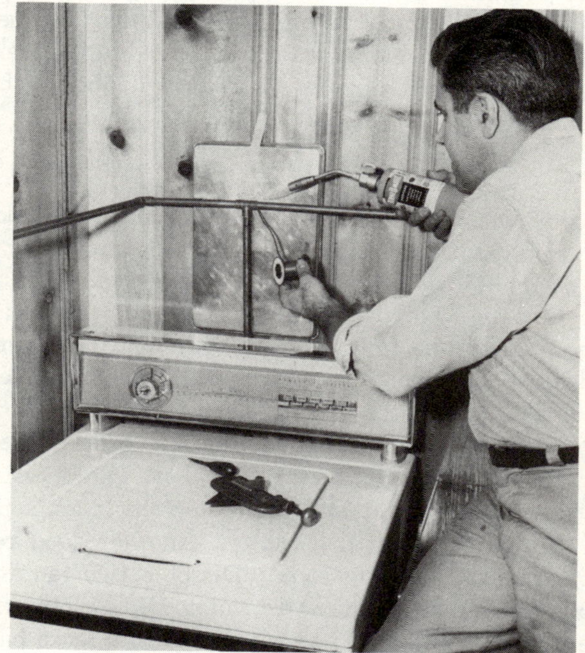

Courtesy Bernzomatic Corp.

Fig. 11-2. Cookie sheet temporarily taped to wall keeps torch heat off paneling.

In second-degree burns, the outer and adjacent layers of skin are killed. The deepest cells of the skin are not affected. The skin will appear red and swollen. There will be blisters, and if broken, they will ooze a liquid. Second-degree burns will heal completely in six weeks or less, but they may leave some scars.

In third-degree burns, all of the skin tissues and some of the underlying flesh are killed. The area may have various appearances: dead white, brown and charred, or bright red. There may not even be swelling at first. The actual depth of the burn cannot be exactly determined for several weeks until the dead tissue has been separated from the live tissue below it. Any third-degree burn area more than about 2 inches in diameter will take months of good care before it heals and, unless early grafting is performed, will show bad scarring.

TREATMENT OF BURNS

First-aid treatment of burns involves two things: reduce the pain and prevent infection. Only first-degree burns can be treated locally. Burns that develop blisters or have the appearance of depth greater than first degree, as described above, should be treated promptly by a physician.

Light Burns

Burns over small areas, showing only red unbroken skin, need very little treatment. Cover the area with a clean cloth or bandage in order to immobilize the movement of the skin of the area and thus prevent pain. The covering should include the area beyond the burned part, and be done in such a way that the covering does not rub against the burned skin. If there is swelling, place clean ice on it or run cold water over it. There are arguments for and against the use of ointments, but one that has proved effective for small burns is *butesin picrate with metaphen.* The butesin picrate is analgesic and anesthetic, and the metaphen is antiseptic. It is available in small tubes at drug stores. Another good remedy is sodium bicarbonate (baking soda). Dissolve two tablespoons of baking soda in a quart of lukewarm water. Dip a cloth pad into it, and place it over the burn. Then cover it with a loose bandage.

Heavy Burns

If the burn is deeper than first degree, the only treatment is to cover the burn with a clean cloth to keep the air off it as much as possible, and call a doctor. Do not attempt to open blisters; only a physician should do this. If the area of the burn appears to be greater than about 15 percent of the body (this is possible in the case of spilled molten solder), consider the possibility of shock, and lay the patient down until the doctor comes. Keep his head and chest lower than the rest of his body. If he can take liquids, give him as much as possible of any suitable liquid but alcohol.

Index

A

Accessory tips, 49
Acid
  abietic, 25
  hydrochloric, 26
Additives, solder, 20
Adhesion, 9
Aids
  desoldering, 126-127
  soldering, 16, 81-82
Alloy (s), 9
  low-melting-point, 8
  tin-lead, 11
Aluminum, 11
  soldering, 22-23
American
  Beauty, 30
  Electrical Heating Company, 30
  Society for Testing Materials, 18
Amine hydrochloride, 26
Antimony, 18, 20
ASTM, 18
Atomic structure, 8
Attachment, drill, 56, 58
Attraction, capillary, 10

B

Barrier, oxide, 10
Base metal, 8, 9
Batteries
  Nicad, 15
  nickel-cadmium, 15
Battery care, 53
Beginners, soldering, 66-69
Blowtorches, gasoline, 32
Board (s)
  printed-circuit, 13, 70-72
  treatment, 104-105
Bottled-gas torch, 114-117
Brazing, 8
Bulb, vacuum, 16
Burns
  classification, 138-140
  heavy, 140
  light, 140
  treatment, 140
Butane torches, 32
Butesin picrate, 140

C

Cable (s)
  connectors, 74-75
  shielded, 75-76
Capillary attraction, 10
Car-battery-operated iron, 45
Char, 27
Charger, 15, 53
Chassis, soldering, 72-73
Chloride, zinc, 26
Chromium, 11
Cleanliness, 27, 64
Cobalt, 11
Coefficient of expansion, 38
Compromise gun, 50-51
Configurations, tip, 13
Connections, solderless, 119-120
Connectors, cable, 74-75
Constant-duty irons, 37
Construction
  details, 37-38
  soldering-iron, 32-33
Contours, tip, 42-43
Control, heat, 38-40
Convection, 9
Copper, 11
  oxide, 24
  print, 130

Cordless
gun, Wen, 60
iron advantages, 52-53
Lenk, 60
Ungar, 54, 56
Weller, 54
soldering iron, 15
Wahl iron, 56, 58
Core (s)
flux, 27-28
inner flux, 22
Corrosion, 18
surface, 10
Costs, iron, 36-37
Creep, 20
Crystal-lattice, 9
Current, electrical, 9

D

Desoldering, 15
irons, vacuum, 133
tips, special, 127-130
Diameters, shank, 15
Different soldering tip, 53
Dip soldering, 96-97
Direct-flame soldering, 116-117
Drill attachment, 56, 58

E

Econopak, 102
Efficiency, heat-transfer, 39
Electrical current, 9
Electrons
free, 8
negative, 9
Electrovert, Inc., 100
Elements, interchangeable heating, 13
Equilibrium, 38
Ersin Multicore, 23
Eutectic
ratio, 19
solder, 18
Expansion, coefficient, 38
Experienced solderers, 69

F

Flux, 8, 10, 11, 63
core (s), 27-28
inner, 22
inorganic, 25-27
liquid, 22
nonrosin organic, 26
organic, 25-27
rosin-base, 12
Forms of solder, 21-22
Free electrons, 8
Friction wheel, 117, 119

G

Gas torch, bottled, 114-117
Gasoline blowtorches, 32
Gold, 11
Graph, tin-lead temperature, 18-19
Gum rosin, water-white, 26
Gun (s)
compromise, 50-51
manufacturers of instant-heat, 48-49
soldering, 15
Wen cordless, 60

H

Hand soldering, 83-86
Heat
control, 38-40
recovery, 40
requirements, 39
sinks, 16
transfer efficiency, 39
Heated tips, torch, 114-116
Heating elements, interchangeable, 13
Heavy burns, 140
Hollis Engineering, Inc., 103
Hydrochloric acid, 26
Hydrochloride, amine, 26
Hygroscopic, 27

I

Induction soldering, 94
Inner flux core, 22
Inorganic flux, 25-27
Inspection, 79-80
Instant-heat
guns, manufacturers, 48-49
iron, 15
Interchangeable heating elements, 13
Ions, positive, 9
Iron (s), 11
car-battery-operated, 45
care, 44-45
clad tip, 42
constant-duty, 37
cordless Wahl, 56, 58
costs, 36-37
details, soldering, 89-90
instant-heat, 15
Lenk cordless, 60
pencil-type, 33-35
right, 62-63
transformer-isolated, 35-36
Ungar cordless, 54, 56
vacuum desoldering, 133
Weller cordless, 54

K

Kester, 23

L

Lead (s), 9
pretinning, 65-66
tinned, 18
Leakage, voltage, 88

Lenk cordless iron, 60
Light burns, 140
Lines, liquidus, 19
Liquid
  flux, 22
  state, 8, 9
Liquidus lines, 19
Low-melting-point alloy, 8

M

Magnesium, 11
Melting
  -point alloy, low, 8
  temperature, 18
Metal (s), 8-9
  atom, 8
  base, 8, 9
  joining, 7-8
  molecule, 9
  special, 122-123
Metaphen, 140
Methods, 121-122
Molecule, metal, 9
Molten, 9
Multicored, 22

N

Neutralize, 27
Nicad batteries, 15
Nichrome wire, 32
Nickel, 11
  cadmium batteries, 15
Noncore soldering, 28-29
Noncorrosive, 25
Nonrosin organic flux, 26

O

Organic
  flux, 25-27
    nonrosin, 26
  resin, 25
  rosin, 25
Oxide
  barrier, 10
  copper, 24
Oxygen, 24

P

Palladium, 11
Pasty
  range, 18
  state, 18
Pencil-type irons, 33-35
Percussive welding, 121
Picks, 16
Pitting, 20
Plastic rosin, 25
Platinum, 11
Point-to-point wiring, 12
  repairs, 124-126
Positive ions, 9
Pot, soldering, 95-96
Precautions, 137-138
Precleaning, 12
Precoating, 21
Preforms, soldering, 95
Pretinning leads, 65-66
Printed-circuit board (s), 13, 70-72
  wiring repairs, 130-132

Q

Qualities, solder, 20-21

R

Range, pasty, 18
Ratio, eutectic, 19
Recovery, heat, 40
Regulation, wattage, 38
Removers, wick solder, 133, 135
Repairs
  point-to-point wiring, 124-126
  printed-circuit board, 130-132
Replaceable tips, 41
Requirements, heat, 39
Resin, organic, 25
Resist, solder, 105-106
Resistance soldering, 8, 90-94
Resoldering, 29
Retinning tip, 43-44
Right
  iron, 62-63
  tip, 62-63
Rosin
  -base, fluxes, 12
  -core solder, 21
  organic, 25
  plastic, 25
  water-white gum, 26

S

Self-regulation, 47-48
Semiconductor (s), 73-74
  devices, 13
Shank diameters, 15
Shielded cables, 75-76
Silver, 11
Sizes, tip, 13, 42-43
Soft solder, 17-18
Solder, 63
  additives, 20
  connection area, 15
  eutectic, 18
  forms, 21-22
  qualities, 20-21
  removers, wick, 133, 135
  resist, 105-106
  soft, 17-18
Solderability tests, 106-110
Solderers, experienced, 69
Soldering, 9-12
  aids, 16, 81-82
  aluminum, 22-23

Soldering—cont
/desoldering aids, 126-127
dip, 96-97
direct-flame, 116-117
for beginners, 66-69
gun, 15
hand, 83-86
induction, 94
iron (s) construction, 32-33
cordless, 15
details, 89-90
noncore, 28-29
pot, 95-96
preforms, 95
representative, 12-16
resistance, 90-94
superiority, 61-62
tip, different, 53
to chassis, 72-73
wave, 97, 100, 102, 103
Solderless connections, 119-120
Solid-shank tips, 41
Solidify, 19
Solidus, 19
Special
desoldering tips, 127-130
metals, 122-123
Splices, wire, 76, 78-79
Stainless steel, 11
Stands, 111
temperature-controlled, 112-113
Steel, 11
Stepped shank tip, 41
Strength, 20
Stripping, wire, 64-65
Structure, atomic, 8
Superiority, soldering, 61-62
Surface corrosion, 10
Switch terminals, 72

T

Tacking, 80
Tarnish, 24
Temperature, 20
controlled stands, 112-113
graph, tin-lead, 18-19
melting, 18
Terminals, switch, 72
Tests, solderability, 106-110
Thermal activity, 9
Tin, 9
lead alloy, 11
temperature graph, 18-19
Tinned leads, 18
Tinning, 21
Tip (s), 41-44
accessory, 49
care, 43-44
configurations, 13
contours, 42-43

Tip (s)—cont
different soldering, 53
iron-clad, 42
replaceable, 41
retinning, 43-44
right, 62-63
sizes, 13, 42-43
solid-shank, 41
special desoldering, 127-130
stepped shank, 41
torch-heated, 114-116
Torch (es)
bottled-gas, 114-117
butane, 32
heated tips, 114-116
Transformer-isolated irons, 35-36
Treatment
board, 104-105
of burns, 140
Tweezers, 16

U

Ungar, 15
cordless iron, 54, 56

V

Vacuum
bulb, 16
desoldering irons, 133
Voltage leakage, 88

W

Wahl Clipper Corp., 56
iron, cordless, 56, 58
Wall, 13
-Lenk, 30
Water-white gum rosin, 26
Wattage regulation, 38
Wave soldering, 97, 100, 102, 103
Welding, 8
percussive, 121
Weller, 15
cordless iron, 54
Wen, 15
cordless gun, 60
Wetting, 10, 20
Wheel, friction, 117, 119
Wick solder removers, 133, 135
Wire
nichrome, 32
splices, 76, 78-79
stripping, 64-65
Wiring
point-to-point, 12
repairs, point-to-point, 124-126
printed-circuit board, 130-132

Z

Zinc, 11
chloride, 26